END OF THE MEGAFAUNA

Sabertooth cat, *Smilodon fatalis*, continental Americas.

END OF THE MEGAFAUNA

THE FATE OF THE WORLD'S HUGEST, FIERCEST, AND STRANGEST ANIMALS

ROSS D. E. MacPHEE

With Illustrations by Peter Schouten

W. W. NORTON & COMPANY
INDEPENDENT PUBLISHERS SINCE 1923
NEW YORK • LONDON

Printed in the United States of America
First Edition

For information about permission to reproduce selections from this book, write to Permissions, W. W. Norton & Company, Inc., 500 Fifth Avenue, New York, NY 10110

For information about special discounts for bulk purchases, please contact W. W. Norton Special Sales at specialsales@wwnorton.com or 800-233-4830

Manufacturing by Worzalla
Book design by Chin-Yee Lai
Production manager: Julia Druskin

Library of Congress Cataloging-in-Publication Data

Names: MacPhee, R. D. E., author. | Schouten, Peter, illustrator.
Title: End of the megafauna : the fate of the world's hugest, fiercest, and strangest animals / Ross D.E. MacPhee ; with Illustrations by Peter Schouten.
Description: First edition. | New York : W.W. Norton & Company, [2019] | Includes bibliographical references and index.
Identifiers: LCCN 2018029730 | ISBN 9780393249293 (hardcover)
Subjects: LCSH: Morphology (Animals) | Body size. | Extinct animals. | Extinction (Biology)
Classification: LCC QL799 .M3227 2019 | DDC 591.4/1—dc23
LC record available at https://lccn.loc.gov/2018029730

W. W. Norton & Company, Inc., 500 Fifth Avenue, New York, N.Y. 10110
www.wwnorton.com

W. W. Norton & Company Ltd., 15 Carlisle Street, London W1D 3BS

2 3 4 5 6 7 8 9 0

FOR CLARE, WHO ALWAYS UNDERSTOOD.

Key-foot glyptodon, *Glyptodon clavipes*, South America. The rufous oven bird (on carapace), *Funarius rufus*, and the buff-necked ibis, *Theristicus caudatus*, are both extant.

[W]e are in an altogether exceptional period of the earth's history. We live in a zoologically impoverished world, from which all the hugest, and fiercest, and strangest forms have recently disappeared. . . . Yet it is surely a marvelous fact, and one that has hardly been sufficiently dwelt upon, this sudden dying out of so many large mammalia, not in one place only but over half the surface of the globe.

—ALFRED RUSSEL WALLACE,

The Geographical Distribution of Animals (1876)

CONTENTS

PREFACE: LOST IN NEAR TIME (x)

1. BIG (1)
2. "THIS SUDDEN DYING OUT" (13)
3. THE WORLD BEFORE US (27)
4. THE HOMININ DIASPORA (43)
5. EXPLAINING NEAR TIME EXTINCTIONS: FIRST ATTEMPTS (67)
6. PAUL MARTIN AND THE PLANET OF DOOM: OVERKILL ASCENDANT (81)
7. ACTION AND REACTION (91)
8. OVERKILL NOW (107)
9. WHERE ARE THE BODIES, AND OTHER OBJECTIONS TO OVERKILL (135)
10. MORE OBJECTIONS: BETRAYAL FROM WITHIN? (149)

11. OTHER IDEAS: THE SEARCH CONTINUES (161)

12. EXTINCTION MATTERS (177)

EPILOGUE: CAN THE MEGAFAUNA LIVE AGAIN? (187)

APPENDIX: DATING NEAR TIME (191)

Glossary (193)

Notes (199)

References Cited (211)

Guide to Additional Reading (223)

Acknowledgments (225)

Credits and Attributions (227)

Index (229)

PREFACE: LOST IN NEAR TIME

Like other environmental hot-button issues familiar to us today—ocean pollution, forest loss, climate change, species endangerment, pathogen proliferation, the plasticpocalypse, and many others—biological extinction is very much in the public eye. This is surely a result of the growing realization, which now seems utterly inescapable, that humans have created, and continue to create, the kinds of circumstances in which species die out well before their time. A few generations ago people were more negligent in their understanding of many things, including our long record of environmental despoliation, because the necessary facts were either unavailable, difficult to acquire and interpret, or frankly irrelevant to their immediate concerns. In the digital age, it is no longer possible to claim ignorance or lack of readily accessible information. Although it may be true that as we

have learned more about human impacts we have come to despair more, the antidote to this is knowledge, for knowledge is the essential basis for informed action.

In this regard, it has often been claimed that the past record of human interactions with the world's biota—the sum total of living things—should have something to tell us about what to expect in future. But this requires that we ask the right questions, not only about what happened in previous times, but also how it happened.

To do this we need to go to the borderland between archeology and paleontology, the sciences that together attempt to document and explain past life in all its abundance and complexity. *End of the Megafauna* is specifically concerned with examining the enigmatic disappearance, in mostly prehistoric but still relatively recent times, of a major fraction of Earth's truly large vertebrates as well as many of their smaller relatives. These losses seem to have occurred quite quickly within any given area, although on a worldwide basis the full process took several tens of thousands of years to finish—if it is in fact finished.

Popular works tend either to assume the cause of these losses is settled or to treat differing viewpoints about causation in a superficial manner. This book takes a different tack, by concentrating on how scientists who study these mysterious disappearances have attempted to explain them. We will be looking at the strengths of their arguments as well as their limitations, because challenge and response is what good science and good scientists are all about. To make discussion worthwhile, some technical concepts have to be introduced, but I have tried to present them in an appetizing manner. Since there is no way I could cover more than a fraction of the scholarship that is relevant to this endeavor, I have purposely chosen to emphasize themes that strike me as the most interesting.

Most paleontologists work in deep time, the distant past, where intervals are measured in millions of years. I study events that, geologically speaking, happened almost yesterday. The losses covered in this volume have been called late Quaternary extinctions, Ice Age extinctions, Late Pleistocene/Holocene extinctions, Anthropocene extinctions, megafaunal extinctions, and modern-era extinctions, depending on when exactly they occurred and what aspects are being emphasized. (Here and throughout, consult the glossary on page 193 for unfamiliar terms.) These labels overlap to a greater or lesser degree, but have different connotations depending on who's doing the defining. The result is potential confusion in the minds of readers. Although I will use some of these alternative wordings from time to time to add variety to my text, I prefer to use the phrase "Near Time extinctions" to encompass all vertebrate losses, however caused, that took place during the past 50,000 years or so. Near Time extinctions as a label has the advantage of being outside the system of formally named intervals that scientists use, yet it embraces the relevant parts of all.

At the same time, Near Time is no more homogeneous than any other period in Earth history, so I take care not to overgeneralize. One distinction I will make throughout is between very late losses, or ones that occurred in the mid-Holocene or later (last 5000 years), and those that occurred appreciably earlier. Scientists are in general agreement that losses in very recent times were fundamentally anthropogenic (that is, human-induced), even if there are uncertainties about what combination of alien species introductions, environmental degradation, and perhaps other factors caused specific disappearances. Although I don't ignore these later extinctions, most of which occurred on islands, my intention is to concentrate on earlier disappearances because there is no consensus regarding their fundamental causation. Compared to very recent losses, the level of uncertainty about what really happened to force these earlier extinctions is much greater, the relevant evidence much thinner, and the explanatory stakes correspondingly higher.

I will mostly focus on the places where I have personally worked because I know them, and their stories, best. I have tried to weave in extinctions that occurred in other contexts, but I acknowledge that I have had to be fairly selective in this, and in any case I have weighed the evidence and analyses in my own way. For a similar reason, I concentrate on what happened to mammals, my special group, although birds and reptiles are certainly part of the story as well and need referencing.

The world of the extinct megafauna is, of course, a lost world, but not one so very different from today's that there is no hope of understanding how these great beasts might have looked and lived. Peter Schouten's exceptionally realistic and beautifully executed paintings provide a luminous window onto this world in all of its extraordinary diversity. Image captions provide a way of talking about the animals themselves, as real, living entities rather than just a depressing series of entries in the thick ledger of Near Time extinctions. Patricia Wynne's maps, charts, and delightful line drawings help to summarize, as well as to throw a spotlight on, critical points that are much better explained in images than words. Very few extinct animals have common names; biologists and paleontologists of course use technical names that nobody has ever heard of. I try to avoid the latter as much as possible in the text, although for completeness proper biological names are provided in the captions to the illustrations.

Any scientific debate that is worth participating in must have its ground rules. For me, one such rule is that all serious positions regarding a chosen topic deserve a hearing, especially when they raise evidential or theoretical issues that deeply affect other such positions. In this book I am much more interested in examining the ins and outs of the various theories we will be exploring than in concluding beyond a reasonable doubt that human overhunting, or climate change, or fireballs, or disease, or some combination of these, or even none of the above, was responsible for this or that Near Time loss. Yet having said this, I recognize that in a debate, after all has been said and done, the audience expects to hear a sense of the house as to which viewpoint has prevailed or has done the best job of covering the facts and answering the majority of arguments pro and con. This we will get to, but first—on to the evidence!

END OF THE MEGAFAUNA

1

BIG

PLATE 1.1. **A GIANT ELEPHANT BIRD** (*Aepyornis maximus*) walks past a pair of extant ringtailed lemurs (*Lemur catta*) in southern Madagascar a thousand years ago. This bird may have weighed as much as 230 kg (500 lb), with eggs up to 1 m (3.1 ft) in circumference. Remains of *Aepyornis* and its close relative *Mullerornis* (plate 4.8) are frequently encountered in subfossil sites, an indication that these birds were probably rather common. Some beaches in Madagascar are littered with the broken eggshells of these massive animals, which strongly implies that they nested there. Elephant birds may have survived long enough to become, in very distorted form, the basis for the legend of the Roc, a wondrous tale carried to Europe by Marco Polo in the early fourteenth century.

Size is of course a relative concept. In biology, body size is usually evaluated in terms of similarities and differences among species in the same general group. For example, the one-celled creatures called foraminifers, which live throughout the world ocean and feed on organic detritus raining down through the water column, are mostly around 1 millimeter (0.04 in) in size. Some species, however, can attain a length of 20 cm (7.9 in). In their microworld, such a relatively enormous size qualifies as megafaunal, which is exactly how these species are categorized by the scientists who study them. In our human-scale world, elephants and great whales qualify in the same way—they are much bigger than other living mammals, which invites questions about why they are that way, and were they always like that, evolutionarily speaking.

To answer such questions, it is first necessary to realize that there is much more, biologically speaking, to being megafaunal[1] than mere massiveness. Depending on the organism, large body size tends to be correlated with a range of physiological and behavioral characteristics that do not necessarily occur in the same way in smaller relatives. As we shall see later on, body size can also be incredibly dynamic, with upsizing or downsizing having taken place in some lineages over remarkably short time scales, especially in island contexts. To mention some simple examples, among land mammals the large body sizes found in bulk-feeding herbivores (think of cows) can be advantageous, since the digestion of plant matter is typically a lengthy, energy-consuming process that can be made more efficient by processing large quantities of food at a time. By contrast, diminutive species, like most rodents, can successfully exploit small, energy-packed foods such as grains, seeds, gums, or insects that larger species generally cannot, or cannot do so efficiently. Naturally, there are always trade-offs: large species may do better in a greater variety of environments than smaller species would, but smaller species tend to have more offspring per birthing cycle. Large-bodied species generally have longer individual life spans than is the rule for small-bodied species, and so on.

Thus the way to think about size is that being big can be just as good an evolutionary strategy as being very small—or really any size in between, for it all depends on circumstances. And, in *almost* all circumstances we know about in the fossil record, being big did not come with an accelerated time stamp, so that large species appeared and then disappeared very quickly. On average, at any given time in the past, large species were no more likely to perish than smaller ones, and most large extinction events show no particular size bias. Among the few exceptions that prove the rule are the very losses we will be considering here.

1. In Quaternary extinction studies, the term "megafaunal" is conventionally applied to species having body sizes greater than 44 kg (100 lb). Although many objections can be leveled against this one-size-fits-all approach to body size definition, it's historically appropriate and will be used throughout this book.

Time and its measurement has to be part of any discussion in paleontology. I have tried to limit my use of technical terms, but some are just unavoidable. Table 1, a simplified version of the geologic time scale, illustrates how major intervals named in this book interrelate. Definitions can be found in the glossary.

TABLE 1 **NAMED INTERVALS USED IN THIS BOOK**[2]

<table>
<tr><th></th><th>EPOCH/STAGE</th><th>GLACIAL/ INTERGLACIAL PERIODS</th><th>RANGE IN YEARS</th><th>INCLUDES</th></tr>
<tr><td rowspan="4">QUATERNARY PERIOD</td><td>Holocene</td><td>Present Interglacial</td><td>11,700 to present</td><td>Modern Era
1500 CE to present

Little Ice Age
1300 to 1850 CE</td></tr>
<tr><td rowspan="2">Late Pleistocene</td><td>Last Glacial</td><td>110,000 to 11,700</td><td rowspan="3">Younger Dryas
12,900 to 11,700

Last Glacial Maximum
23,000 to 27,000

(Near Time 50,000 to present)</td></tr>
<tr><td>Last Interglacial</td><td>130,000 to 110,000</td></tr>
<tr><td>Middle & Early Pleistocene</td><td>At least 4 more glacial & interglacial cycles</td><td>2.6 million to 130,000</td></tr>
<tr><td></td><td>Late Pliocene</td><td>Pre-Pleistocene Glacial</td><td>3.6 to 2.6 million</td><td>Greenland ice sheet forms 3 million years ago</td></tr>
</table>

2. Cell heights not to scale. Epoch preceding Quaternary is Pliocene (5.3 to 2.6 million years ago).

The American Museum of Natural History in New York City is world famous for its vertebrate paleontology halls, where the story of vertebrate life is traced from its late beginnings to the near present, as told by the most direct form of evidence we have, the fossils themselves. Visitors naturally want to see the dinosaurs, but there are many other extinct animals on display, including those presented in the Lila Acheson Wallace Wing of Mammals and Their Extinct Relatives. Like the other fossil halls, this one presents mounted

FIG. 1.1. **ICE AGE PROBOSCIDEANS:** *Mammuthus columbianus*, the Columbian mammoth (*left*) and *Mammut americanum*, the American mastodon, two iconic North American megafauna.

skeletons of vanished beasts, prodigious in number and variety. The difference here is that many of the species seem quite familiar, not only because of frequent cameo appearances in feature-length films, documentaries, and cartoons, but also because they have reasonably

MAMMOTH ON THE MENU

When people hear that I am a Quaternary paleontologist who works in places like Siberia and Yukon, I am sometimes asked rather startling questions, such as:

QUESTIONER: "HAVE YOU EVER FOUND A MAMMOTH MUMMY?"
MY ANSWER: "YES, I HAVE."
FOLLOW-UP QUESTION: "DID YOU EAT IT?"

Whether I say yes or no there is always a sharp intake of breath and the inevitable further inquiry, "What did it taste like?" or, alternatively, "Why not?"

To set the record straight, I have not, ever, eaten the flesh of a mammoth or any other beast encountered on the frozen tundra. The reason is easy to relate in the form of another question, "Would you eat that road pizza of a deer you saw at the side of the interstate last week?" Most of my readers would, I assume, say no, on the excellent ground that any such pizza so fortuitously encountered might not be a good dietary choice, for many reasons. I have seen mammoth skulls that were literally bursting with larval stages of Pleistocene flies and beetles, obviously less fastidious diners than I. This was recycling, Quaternary style, and a very good thing too for the ecosystem at the time it took place.

Despite what you see in the movies, tundra mummies weren't flash frozen or suddenly covered by advancing ice. Instead, they simply died, from whatever cause, and rotted in place while drifting sediments covered them, only to be discovered by people—or their dogs—many millennia later.

Although upon discovery the carcasses of mammoths were often said to be as fresh as well-frozen beef or horsemeat, upon excavation they usually began to smell horribly and only dogs showed any appetite for them. In this 1866 effort to portray the Adams Mammoth in life (see figure 5.3), the ears are far too large, the tusks curl the wrong way, and the lengthy mohawk running down the animal's back is sheer invention. Correcting such errors had to await truly scientific study of permafrost mummies.

FIG. 1.2. **GIANT GROUND SLOTH:** *Lestodon armatus* (South America).

close living relatives. They look *right*, or at least some of them do, as though they could be living somewhere today. This is where we are going to pause for a while, because what's on display here is exactly what this book is all about.

Take the duo dominating one end of the wing, a Columbian mammoth and an American mastodon (see figure 1.1). Both are very definitely proboscidean, or elephantlike, in body form, although their last common ancestor lived about 25 million years ago. On the North American mainland, populations of mammoths and mastodons were still living as recently as 12,000 years ago; all were gone 1,000 or so years later. A couple of island-bound groups of woolly mammoths struggled on, but these too had disappeared by 4200 years ago. To put this in the perspective of human history, that's when the Middle Kingdom of ancient Egypt and the Caral civilization of pre-Columbian Peru were flourishing. Humans, of course, persisted; so did Asian and African elephants. These magnificent beasts didn't (see box, page 6). Why?

Elsewhere in the hall we see members of Xenarthra, today an almost exclusively South American group that includes living armadillos, tree sloths, anteaters, and their extinct close relatives. The largest of the living xenarthrans, the giant anteater (*Myrmecophaga tridactyla*), may weigh as much as 40 kg (88 lb), but as late as 12,000 to 13,000 years ago there were several xenarthran species in both North and South America that may have weighed as much as 2000–4000 kg (4200–8800 lb). Among these was gigantic *Lestodon* (see figure 1.2), probably an inoffensive herbivore despite its huge claws and looming posture. Its closest living relatives are the two- and three-toed tree sloths, *Choloepus* and *Bradypus*, no species of which weighs more than 5 kg (11 lb). They made it, *Lestodon* didn't. Why?

And what would a visit to the Pleistocene be without a chance to see a long-dead hypercarnivore, albeit at a safe distance? For genuinely menacing looks, nothing compares to the sabertooth cat (*Smilodon fatalis*) (see figure 1.3). Lionlike in build and doubtless in strength, but extravagantly armed with massive, daggerlike canines, sabertooths differed radically from our surviving large carnivores. How the sabertooth's teeth functioned in hunting continues to puzzle scientists, more than a century and a half after its initial description. However, those supersized canines must have performed a useful duty of some kind, because a similar adaptation appeared during the evolution of several other, quite unrelated carnivorous mammals. But if this was such a successful design, why is it not seen today?

In addition to these well-known giants there are many, many other Quaternary species known only to specialists, such as the fierce-looking but presumably entirely inoffensive giant koala lemur depicted in figure 1.4. There were also flightless birds three times the size of an ostrich, other big lemurs the size of gorillas, and lizards that weighed half a ton (see plates 1.1, 2.1, and 2.2, respectively). These amazing creatures prospered in their native

FIG. 1.3. **SABERTOOTH CAT:** *Smilodon fatalis* (North and South America).

environments for hundreds of thousands of years or more without suffering any imperiling losses. But beginning about 50,000 years ago, something started happening to large animals here and there, and eventually nearly everywhere over the planet. Depending on their contexts, they sometimes disappeared singly, at other times in droves. Other species were affected as well, but it is the losses at the large end of the spectrum of body sizes that immediately attract our attention. Size must have mattered, because their smaller close relatives mostly weathered the extinction storm and are still with us.

So why did these megafaunal extinctions occur?

A short but appropriate reply would be that there is no satisfactory answer to this question—not yet, not in the sense that there is widespread agreement among those who, like me, have occupied themselves with addressing it as well as they can from a scientific perspective. Just because past events lie close to the present does not mean their interpretation is any easier or the evidence associated with them more tractable than ones more distant in time. Despite the immense number of things that we actually do know about the disappeared, we continue to rely on intuitions, hunches, hints, indications, possibilities, and half-baked ideas to explain their loss. Some researchers think they have the answer, at least for some of the departed; others are not so sure. The debate continues as fresh leads are traced and dead ends abandoned or refashioned in order to accommodate new evidence. It is a great time to be a Quaternary paleontologist!

FIG. 1.4. **GIANT KOALA LEMUR:** *Megaladapis [Peloriadapis] edwardsi* (Madagascar).

2

"THIS SUDDEN DYING OUT"

PLATE 2.1. **THE GIANT SLOTH LEMUR** (*Archaeoindris fontoynonti*) was the largest of the Malagasy primates. It may have weighed as much as 160 kg (350 lb), which is the size of a big male gorilla. Few remains of this heavyweight species have been found, so many details concerning its adaptations remain obscure. Although *Archaeoindris* possessed long arms like its smaller relative *Palaeopropithecus* (plate 7.6), whether it was an active arborealist seems doubtful. If it was, it must have "walked" through the trees, carefully repositioning one limb at a time as do male orangutans today, and for the same reason—despite having long arms and grappling-hook hands, they are much too big for rapid, suspensory locomotion. The small extant lemur at the right is the aptly named bamboo lemur (*Hapalemur griseus*), which comes in at a modest 2.5 kg (5.5 lb).

THE QUESTION OF CAUSE

Everything about the world we live in today—its geography, climate, biotas, even ourselves—was shaped and reshaped multiple times during the last 2.6 million years (see table 1, page 3, and glossary). This was the world of the Quaternary megafauna (e.g., plates 2.1 to 2.4), the "hugest, and fiercest, and strangest" members of which mysteriously disappeared near the end of this interval, just as Charles Darwin's contemporary and fellow evolutionist Alfred Russel Wallace said they did in the epigraph that opens this book. Considered as a group, and compared to other major loss events in the geological record, Near Time extinctions present several unusual features.

First, they had a peculiar distribution in both space and time: they occurred on most continents as well as many islands, but at very different points in different places. Although some events had impacts over wider areas than others, there was no simultaneous, planetwide crash as there had been in some of the mass extinctions of the distant past. Some losses seem to have been brutally quick, taking place over comparatively short periods, from a few decades to a few centuries. In other cases, the speed of extinction is more uncertain, or they simply took much longer for no obvious reason. Importantly, wherever and whenever they occurred, the number of disappeared was well beyond any conceivable background rate of extinction. (Background rate means the slow, entirely natural drip, drip, drip of isolated losses as individual species, like individual lives, come to the end of their possibilities.)

Second, terrestrial vertebrates were exclusively affected, or so we think. Marine vertebrates largely escaped extinction until the modern era (the most recent 500 years of Near Time), suggesting that, for most of the past 50,000 years, the agent of extinction that brought destruction to those living on the land did not similarly spell disaster for those in the sea. Nonvertebrate casualties during this interval are almost unknown, but it's an open question whether that reflects fact or just our inattention to the fate of the askeletal and less charismatic among us.

Third, there is again the curious matter of size, for the most bizarre feature of Near Time extinctions as a group is that, within individual faunas, it was usually the largest species that were the most seriously affected (see fig. 2.1). It is for this reason that they are often termed "megafaunal" extinctions, even though not all large species came to grief, and smaller species were also impacted to at least some degree wherever Near Time extinctions occurred. I have no objection to this usage, any more than I have an objection to people calling the Fifth Mass Extinction (also known as the K/Pg Mass Extinction) the "dinosaur extinctions," despite the fact that nonavian dinosaurs made up only a small fraction of all the species that came to grief 66 million years ago. In Near Time, size mattered a lot, and it remains a powerful predictor of who lived and who didn't in faunas across the planet. But

FIG. 2.1. **VANISHED (AND VANISHING) GIANTS:** During the past 50,000 years, large mammals lived nearly everywhere habitable on Earth, including many of the world's islands. Most of them are now extinct; some representative examples are depicted on the map as outline drawings. Except for living rhinos and elephants (shaded drawings), all truly large or supermegafaunal species (over 2000 kg/4400 lb in body size) have completely disappeared.

it is actually low reproductive rates and slow maturation—shared physiological traits of most large mammals—that make large species especially prone to extinction under certain circumstances, not their large size per se. Anyway, as a descriptive phrase, "the low reproductive rate extinctions" definitely lacks the punch of "megafaunal extinctions," so I'll stick with this familiar labeling.

Of course, nonmarine vertebrates of large body size are still with us—think of elephants, ostriches, and Komodo dragons. But compared to 50,000 years ago, when I start the Near Time extinction clock, there aren't many big ones left. In North America at present, for example, there are about a dozen species that qualify as megafaunal; before the extinctions began there were at least thirty more. Similar large-scale continental losses occurred in South America and Australia, but intriguingly not in Africa or southern Asia, where losses

PLATE 2.2 (PREVIOUS PAGE). **AN ENORMOUS GOANNA** (*Varanus [Megalania] priscus*) lunges at a wallaby in the Naracoorte region of South Australia 50,000 years ago (plate 7.4). The goanna is the largest known Quaternary lizard; its smaller relative, the highly carnivorous Komodo dragon (*Varanus komodensis*) of Flores and nearby islands, is the largest living today. Fossils of the goanna are uncommon and there are no substantially complete skeletons. This is not necessarily unusual for top predators, which tend to have naturally low population sizes. However, enough is known to hazard some reasonable guesses about its vital statistics. Midrange size estimates hover around a length of 5–6 m (16–19 ft) and a weight of 300–500 kg (660–1100 lb); by contrast, adult Komodo dragons weigh only 70–90 kg (150–200 lb).

were fewer and staggered over considerable intervals. The loss picture for northern Eurasia lies in between these extremes. Only a handful of oceanic islands (i.e., ones far distant from continents) supported truly megafaunal mammals and birds. None of the biggest ones have survived, although a few large island reptiles (mainly crocodiles and tortoises) have managed to hold on here and there.

These, then, are the predominant pattern features of Near Time extinctions, and they need explaining. Two quite different schools of thought have tried to do this. The first invokes climate change as the principal driver of losses; the second, human destructiveness of all kinds. As we shall see, there are many gradations in the content of arguments supporting these positions, but there are likewise some key differences that deserve initial emphasis. Climate change arguments tend to discount any important role for people, and regard Near Time extinctions, especially the ones that occurred on continents, as the regrettable consequence of the alignment of natural forces. Human impact arguments discount environmental changes, even rather dramatic ones, as having had little or nothing to do with the majority of losses, which were instead due to the kinds of depredations that humans seem particularly adept at committing. The implied corollary is that, absent deleterious human activities, the world would still possess hundreds of species that we have cheated out of their full tenure (see box, page 19). In addition, there are intermediate or hybrid positions, as well as quite novel takes on the evidence that invoke other factors, as we shall see.

Climate change arguments are about nature turning on itself, impassively destroying life like some careless child breaking its toys, then moving on, oblivious. People, if present in a specific place when the carnage began, were mere observers or even among those affected, and were certainly not the prime cause. With human persecution as agent the roles are reversed or, better, turned inside out. We become the careless child, killing indiscriminately in the timeless garden of nature. Yet between these two poles of interpretation lies a vast evidential chasm filled with uncertainties. Why did Near Time losses happen at all? Why is it so difficult to identify causal links? What lessons, if any, are to be drawn from Near Time extinctions for projected future losses?

HOW FAST CAN A SPECIES DECLINE?

There is, of course, no rule about how quickly a population can dwindle away. However, the fastest decline on record for a vertebrate with an initially large population size is that of the passenger pigeon (*Ectopistes migratorius*). This species plummeted from literally billions in the nineteenth century to zero by the early twentieth century. Although many theories have been proposed to explain the bird's rapid decline, overhunting is the most supported.

AUTUMN 1813

1,115,136,000

The number of passenger pigeons in a single flock in that year, as estimated by John James Audubon during a day's travel along the Ohio River.

AUTUMN 1913

1

The number of living passenger pigeons known to still be in existence in that year. The last of its kind died a year later.

"Martha," the last recorded passenger pigeon, died at the Cincinnati Zoo on September 1, 1914. Natural history artist Patricia Wynne commemorated the hundredth anniversary of her passing by drawing this portrait on that date in 2014. She now rests in the collections of the Smithsonian Institution.

PLATE 2.3. **THE GIANT IRISH DEER** (*Megaloceros giganteus*) was as large as a moose but it is chiefly famous for its extraordinary antlers. Although its common name may suggest it was limited to Ireland, this is not the case: its Pleistocene range was enormous, stretching from western Europe to China. And, for a time, it was a survivor—at least one population, living in Russia's Ural Mountains, managed to survive until about 7700 years ago, long after the end of the Pleistocene. The effects of climate change, followed by slow population collapse, have been offered as an explanation for its final demise, although as usual the evidence to support this inference is scanty. The giant Irish deer probably fed on a variety of nutritious plants, including forbs and shoots as well as browse in areas where forest existed. The bird foraging in the vegetation just uprooted by the deer is the extant ring-necked pheasant (*Phasianus colchicus*).

AN IDEA, BUT NOT YET AN ANSWER

As a graduate student at the University of Alberta in the 1970s, I was introduced to a troubling idea, one that hasn't left me alone since, even though I have spent most of my scientific career on other topics. The idea was that the losses that occurred in Near Time may be causally unique within Earth history. This may not sound very revolutionary when stated in the bland and careful language of science, but it is. Extinction is a natural process and, as such, is assumed to have perfectly natural causes. The individual factors thought to have induced the mass extinctions of the distant past, such as extraterrestrial impactors, oxygen depletion in the ocean, or large-scale volcanism, may be awful to contemplate because of their scale. But in the end they all ultimately stem from the operation of processes to which Earth has been subject since its formation—ordinary occurrences, carried to extraordinary extremes.

Paul S. Martin had a different solution in mind for very recent losses, and it was this that has given me no peace.

Beginning in the 1960s, Martin claimed in a series of influential papers and edited volumes that, in terms of mechanism, the extinctions of the recent past were nothing like those of more ancient times, because they wouldn't have happened in the way they did without ourselves, *Homo sapiens*, unleashing a megafaunal Armageddon. He pointed out that when groups of people just like us began spreading across distant parts of the planet, they would have repeatedly encountered species that had never previously experienced modern humans. These animals wouldn't have initially recognized the mortal danger represented by these awkward-looking, slow-moving, never-before-encountered upright featherless bipeds, casting hungry looks in their direction.

Martin made a provocative connection between prey naïveté, as it is technically called, and opportunity: if the newly arrived humans weren't already big-game hunters, they became such overnight, sensing that they could hunt clueless megafauna at unprecedented levels and benefit themselves in the way that humans usually do, by increasing their population sizes. Thanks to overhunting or overkilling (essentially synonymous terms, one emphasizing the mechanism, the other the result), Martin argued, the carnage was so great that species in hunters' crosshairs never had a chance to recover after the initial onslaught. Although he allowed that extinctions occurring at different times in different places would have had their own distinctive features, he strongly insisted that common factors must have been in operation to explain why it was always such a bad deal to meet up with humans for the first time.

According to Martin, a further indication that people were implicated in these losses was the fact that the largest species had much higher rates of loss than smaller ones did. For example, well over one hundred mammal species, three-quarters of which were larger than 100 lb (44 kg), disappeared in North and South America combined at the end of the

PLATE 2.4 (PREVIOUS PAGE). **A GIGANTIC SHORT-FACED BEAR** (*Pararctotherium pamparum*) ambles over the pampas in southern South America in the Late Pleistocene. Large examples of this bear may have weighed as much as 1500–1750 kg (3300–3800 lb), which is up to three times the size of the largest polar bear on record. Pleistocene species of short-faced or tremarctine bears occurred in both North and South America (plate 11.4). All are now gone except for the very modestly sized (100 kg/220 lb) spectacled bear *Tremarctos ornatus*, which is mostly confined to upland habitats in South America. Giant bears may not have been especially carnivorous, relying instead on a wide variety of foods just as many bears do today. During the Late Pleistocene, the grasslands of South America would have been much more extensive than at present, as well as both drier and cooler (figure 3.5). The glyptodon with the distinctively flattened shell is *Sclerocalyptus heusseri* and the llamas are members of the extinct species *Hemiauchenia paradoxa*; both seem to have preferred semiarid conditions.

Pleistocene. So did many birds, including the largest raptors on these continents, as well as some reptiles. At about the same time, disappearances in northern Eurasia added another dozen true megafauna to the extinction list, as did those occurring somewhat earlier in Australia. The same thing happened on islands the world over when humans finally got to them, although in these cases the losses were even more thoroughgoing, with species of all sizes disappearing, up to and including entire faunas.

The total wastage worldwide is uncertain and at least partly dependent on one's concept of what a species is and how species boundaries are detected in nature. However, I think a plausible estimate would be on the order of 750 to 1000 vertebrate species, if all known losses between 50,000 years ago and the present were included. There are surely many more extinct species that died out recently but are not yet formally named or lie undiscovered in museum drawers awaiting proper study. In Martin's view, it was the sheer number of repetitions of this phenomenon—in which the animals disappeared without regard to adaptations, climate, or environment, and only after the arrival of the first humans—that made his overkill scenario so compelling. Nor was it necessary to believe that humans killed off every species that disappeared. Large mammalian carnivores and scavenging birds, for example, would have been deleteriously influenced by the loss of prey species, such as large herding ungulates, that the human hunters preferentially sought and quickly exterminated.

I found Martin's argument intriguing, but its likelihood seemed hard to assess. I did not come back to the topic of Near Time extinctions until I began to run my own field expeditions in Madagascar, the West Indies (here grouped as Antillea), and, eventually, Siberia and northern Canada. That required making excavations and collecting remains of former inhabitants. In doing so, I encountered evidence of exactly what Martin had been talking about—species no longer extant whose existence seemed to have been sharply truncated in very recent times. The thing that I could not get my head around—then or now—was what had really happened. Were these extraordinary losses repeatedly due to overhunting, as Martin claimed; to some form of ecological collapse driven by other forces, as others argued; or to something else entirely, as yet unimagined? Could Martin's argument actually be right? To properly address such questions, in following chapters we'll gain some needed background by briefly looking at some of the major transformations that came to define the world of Near Time.

3

THE WORLD BEFORE US

PLATE 3.1. **GIANT GROUND SLOTH** *(Lestodon armatus)*: Sloth species of the size of *Lestodon* (2000 kg/4400 lb; see figure 1.2) were latecomers in sloth evolution, perhaps because niches opened up in the late Cenozoic that were unavailable earlier. In any case, during favorable parts of the Pleistocene, ground sloths of different kinds were able to exploit habitats ranging from Yukon to Tierra del Fuego, from dry grasslands to the Andes. Given their size they were unlikely to have lived in dense, closed forest. *Lestodon* is known mostly from the southern cone of South America, but fossils have also been found in Venezuela, suggesting that at times this huge animal was able to range across much of the continent. *Lestodon* was undoubtedly a bulk feeder, but because it had adaptations that most other large South American herbivores lacked, such as powerful claws and possibly the ability to feed in a bipedal posture, it could have subsisted on anything within its 2 m (6.6 ft) reach, from roots and tubers to grasses and forbs to fruits and tree leaves. The bird searching for external parasites is the extant smooth-billed ani (*Crotophaga ani*).

LATE QUATERNARY CLIMATE AND CLIMATE CHANGE

By acting as a powerful agent of natural selection, climate must always have played an influential role in shaping Earth's biodiversity. But as an explanatory device, climate *change* presents a problem: if climate is always in flux, how extraordinary does a change have to be in order to cause species losses, particularly a lot of them all at once? This is actually a profoundly difficult question to answer, as the number of possible variables to consider is effectively limitless. Let's look at some of the more accessible ones.

The first thing to appreciate about the Quaternary in comparison to other portions of the Cenozoic (the geological interval which includes the last 66 million years) is that it has been relatively cool. The epoch immediately preceding the Pleistocene, the Pliocene (5.3–2.6 million years ago), was appreciably warmer, with an ice-free Arctic and boreal (conifer-dominated) forests stretching as far north as northern Greenland and the Arctic Archipelago north of mainland Canada. Then, about 3.3 million years ago, a major cooling trend began, signifying that something fundamental was happening to the Earth's climatic regulation. Many causes for the changeover have been proposed: alteration of major oceanic thermohaline circulation patterns due to the final closing of the Panamanian isthmus, reduction of greenhouse gases, effects of long-term cycles in the Earth's position and orientation relative to the Sun (known as Milankovitch cycles), or perhaps some combination of these factors and others not yet recognized. In the Northern Hemisphere, the late Pliocene ice sheets reached deep into North America, Greenland, and Eurasia. After several hundred thousand years, the cold trend reversed and the Earth was warmed by a long interglacial episode, only to plunge back into another ice age thousands of years later. Things continued in this manner through the subsequent Pleistocene epoch, cold and warm intervals alternating back and forth, right up to the warm Holocene in which we are living today. However, although these cycles were repeating, they were not uniform in either their length or constancy of conditions. Overall, the best way to think about climate change in the interval we will be concentrating on, the late Quaternary (the last 130,000 years), is to compare it to a global-scale roller coaster—full of leaps and dives, some sharp, others gentle, and surely harrowing for the unprepared (see figures 3.1, 3.2, and 3.6).

EARTH IN THE ICE AGE

A good way to appreciate the difference between Then and Now is to take a global snapshot of the Earth when climate was the opposite of what it is today—in the heart of the last ice age. (This is just an overview, emphasizing conditions on northern continents; details relevant to extinction biology in specific areas will be introduced in later chapters.)

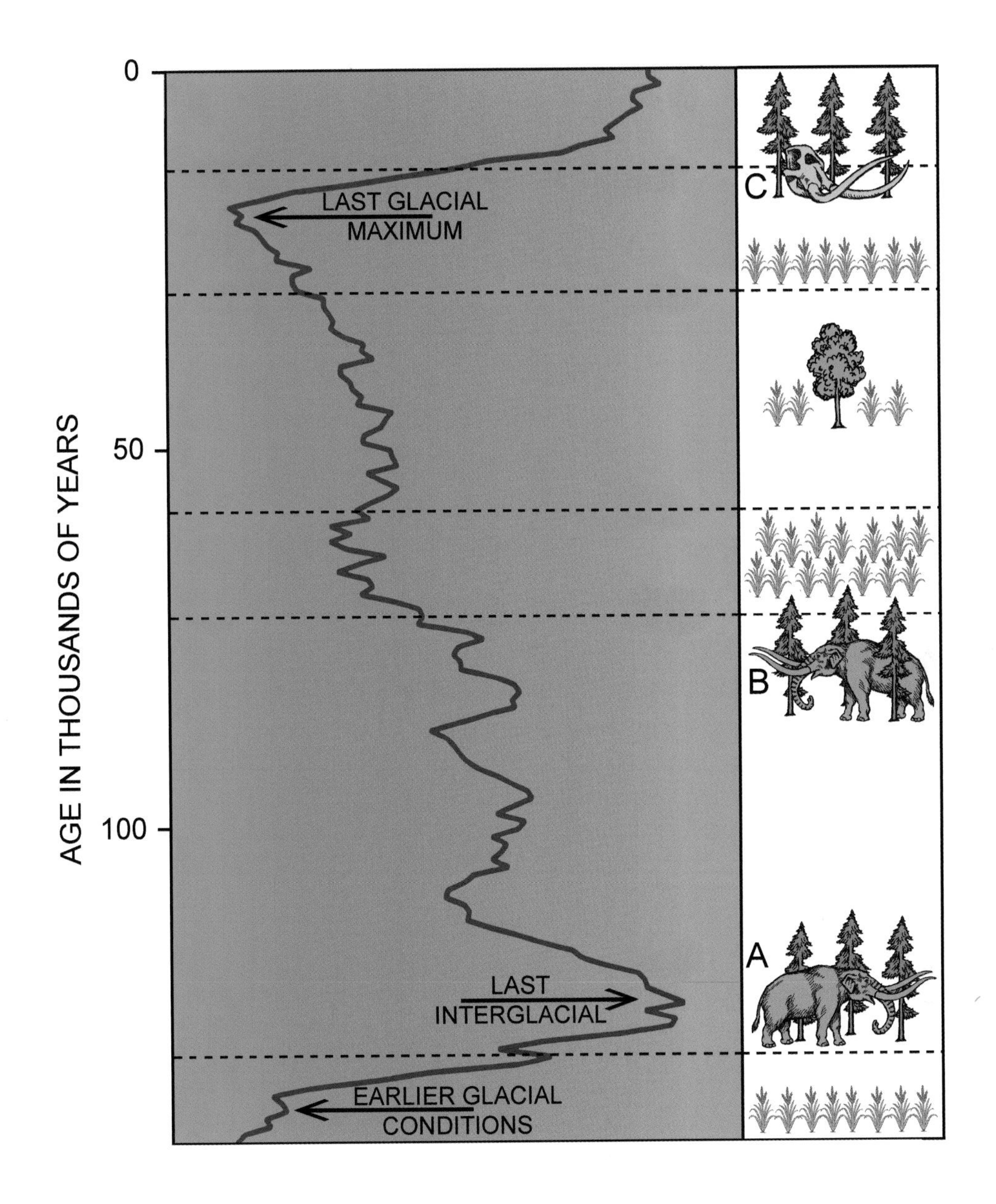

FIG. 3.1. **THE LATE PLEISTOCENE, 130,000 YEARS OF ROLLER-COASTER CLIMATE CHANGE:** The graph line tracks mean annual temperature change over the course of the Late Pleistocene: relatively cool times are indicated by teal tones, warmer times by red. Also indicated are the maximum warm and cool periods during this interval, the Last Glacial Maximum (23,000 to 27,000 years ago), and the Last Interglacial or Interglaciation (123,000 to 130,000 years ago). These changes had profound effects on the flora and fauna. For example, mastodons (right-hand column) were present in midlatitude North America throughout the Late Pleistocene until their complete extinction about 11,700 years ago (*A–C*). But in Yukon and Alaska they disappeared much earlier (*B*), probably as conditions began to shift toward colder temperatures 60,000 to 70,000 years ago and boreal forest (*dark green plants*), their preferred habitat, was replaced by tundra vegetation (*light green plants*).

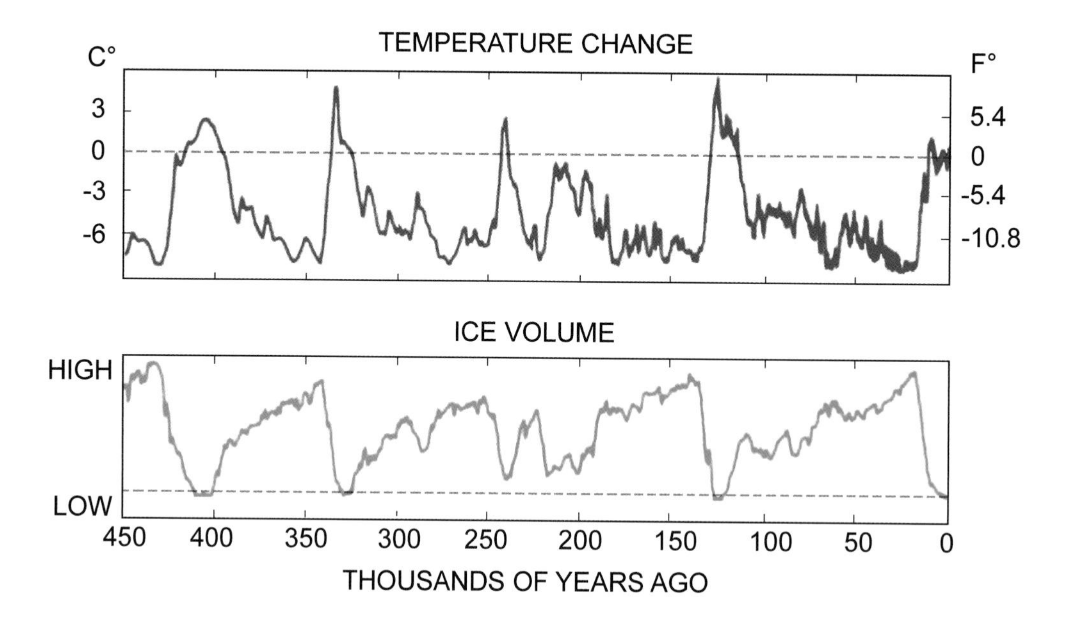

FIG. 3.2. **TEMPERATURE AND ICE VOLUME:** There has been strong periodicity in ice advance and retreat during the Quaternary, which is in turn correlated with temperature. For the past 800,000 years, major advance/retreat intervals have occurred—very roughly—on a 100,000-year cycle. Note that temperatures have generally been cooler than the present (current averages represented by the dashed line); only 2 percent of Pleistocene time has been as warm as or warmer than now.

In the coldest part of the last glaciation 23,000 to 27,000 years ago, technically known as the Last Glacial Maximum, a large portion of the Northern Hemisphere was covered by ice sheets and montane glaciers (see figure 3.3). North America was especially affected, with its northern half entirely covered by three coalescent ice sheets: the Laurentide, centered on Hudson Bay and sprawling outward from there to cover all of eastern and central Canada and the contiguous parts of the United States; the Innuitan, centered on the islands of the Arctic Archipelago, continuous with the Greenland ice sheet; and the Cordilleran, covering all of British Columbia and southern Alaska eastward to (and including) the Rockies. Northwestern Eurasia, including Scandinavia and much of the European part of northern Russia, was also under ice. Smaller sheets dotted mountainous landscapes in many parts of the world—the Alps, Caucasus, even Mount Kenya in Africa.

Despite the terminology, these ice "sheets" would not have been flat and featureless. They would have varied in thickness and topography, depending on precipitation levels and the lay of the land over which they were draped. Also, because ice sheets locked up so

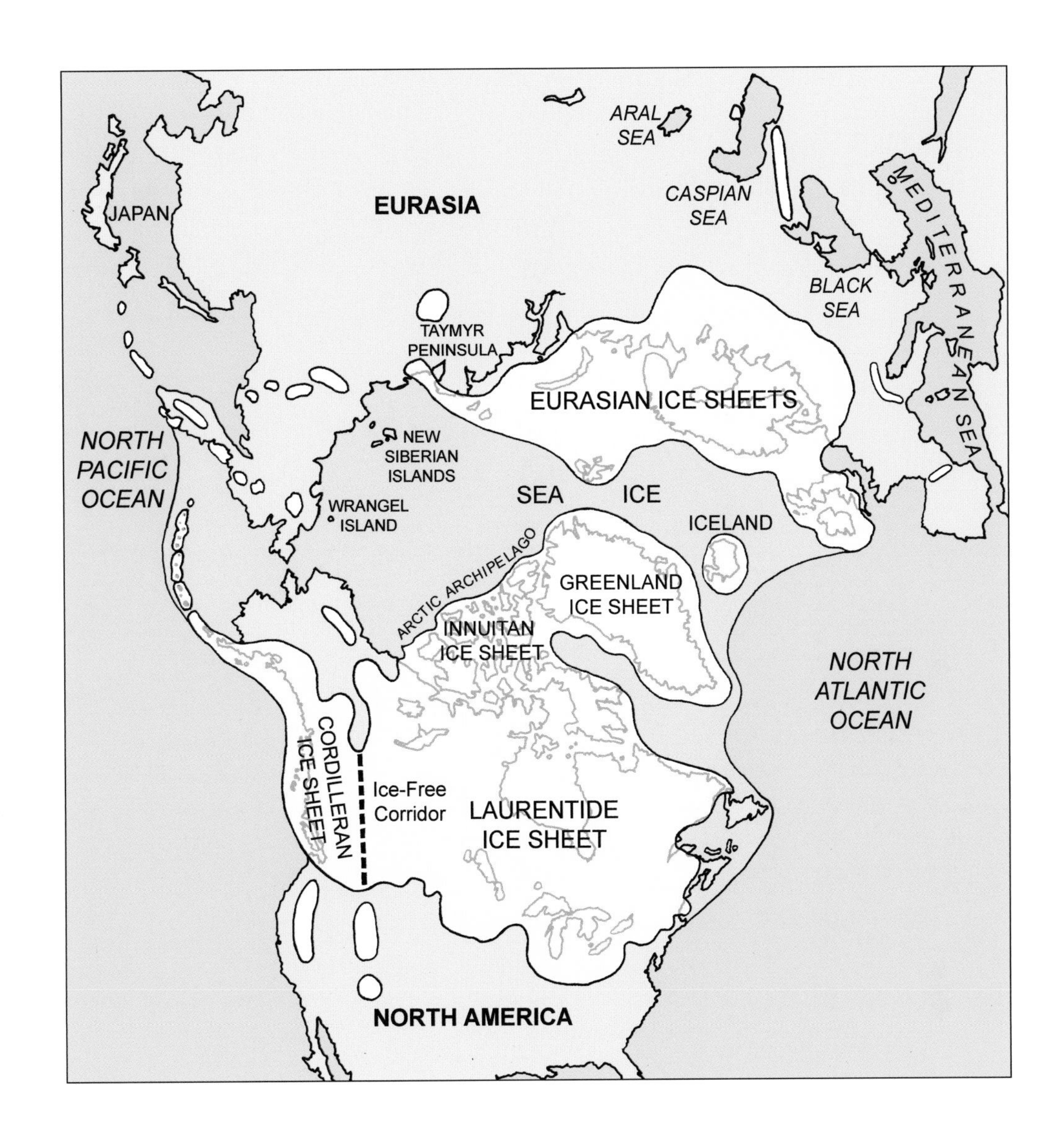

FIG. 3.3. **LAST GLACIAL MAXIMUM, DISTRIBUTION OF ICE SHEETS IN NORTHERN HEMISPHERE:** Large parts of Eurasia and North America were repeatedly covered by ice sheets during the Pleistocene. This map, which shows modern coastlines for ease of reference, displays maximum distribution of ice approximately 25,000 years ago. The ice-free corridor between the Cordilleran and Laurentide ice sheets, closed at this time, would not begin to reopen until about 14,000 years ago. (For the effect of ice sheets on sea level, see figure 3.4.)

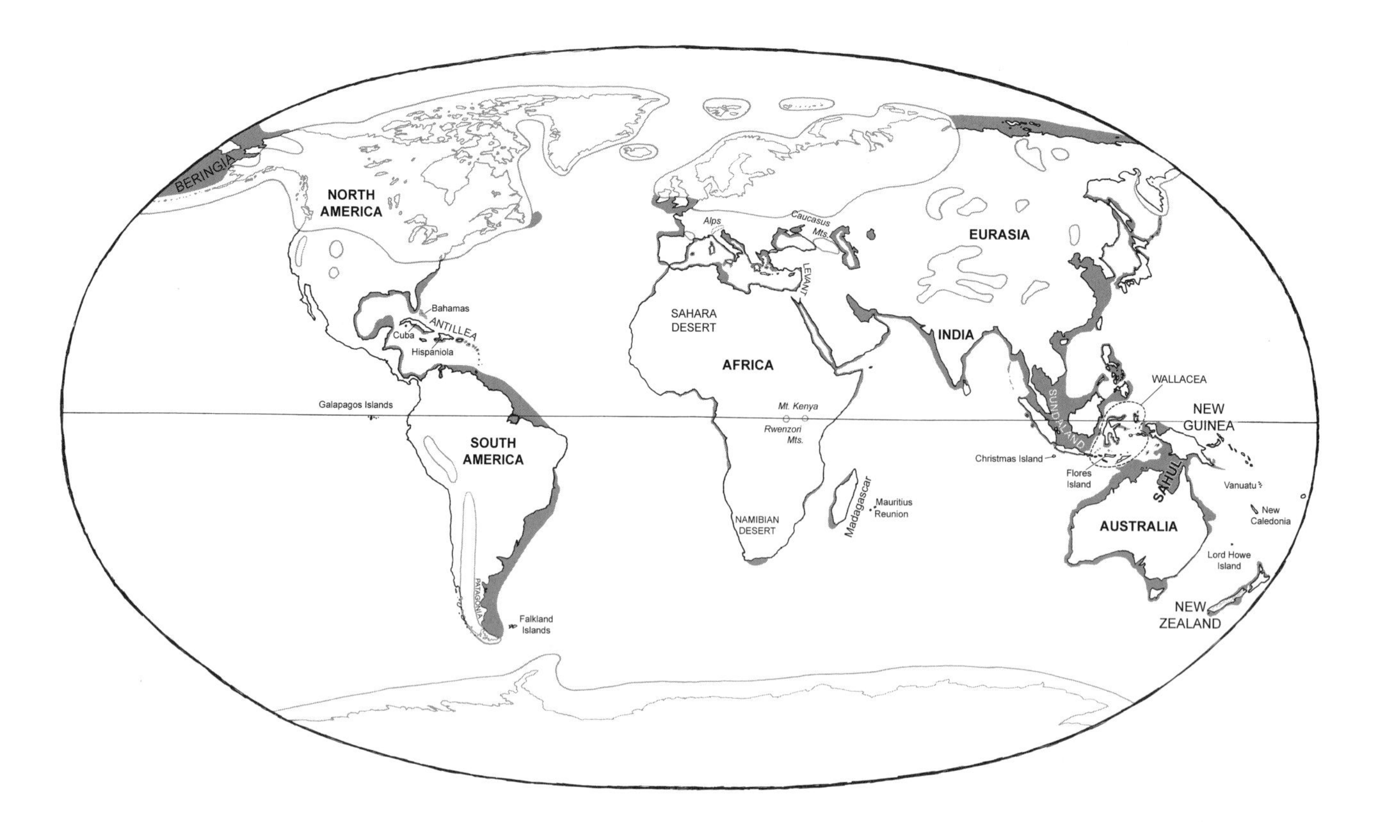

FIG. 3.4. **LAST GLACIAL MAXIMUM, EFFECTS ON SEA LEVEL AND COASTLINES:** Vast areas of shallow sea floor adjacent to continents and many islands were exposed due to the growth of terrestrial ice sheets and consequent removal of water from the oceans. Exposure is most evident in Beringia and southeast Asia, but nearly everywhere was affected to some degree.

much of the world's water, lowered sea level resulted in exposure of continental shelves (and smaller insular shelves) all over the world (see figure 3.4). A good example is the Bering land bridge between easternmost Asia and Alaska, which formed when a huge area of sea floor was exposed between the two continents. As an ecological zone, Beringia, as it is known, extended from the Taymyr Peninsula in central Siberia to the interior of Yukon. It would have supported a specific mix of grasses, low bushes, and other herbaceous plants (forbs) known as the mammoth steppe. Similar vegetation would have grown in the periglacial regions along the southern border of the ice sheets.

Eurasia, as the largest continent, exhibited the greatest variety of ecological conditions. Although much of the mid- and high-latitude parts of the continent was ice free at this time, this was because of a relative lack of precipitation, not because it was warmer than northern North America. Indeed, virtually all of northern Eurasia was occupied by either extreme polar desert or treeless steppe. At lower latitudes, deserts occupying the Middle

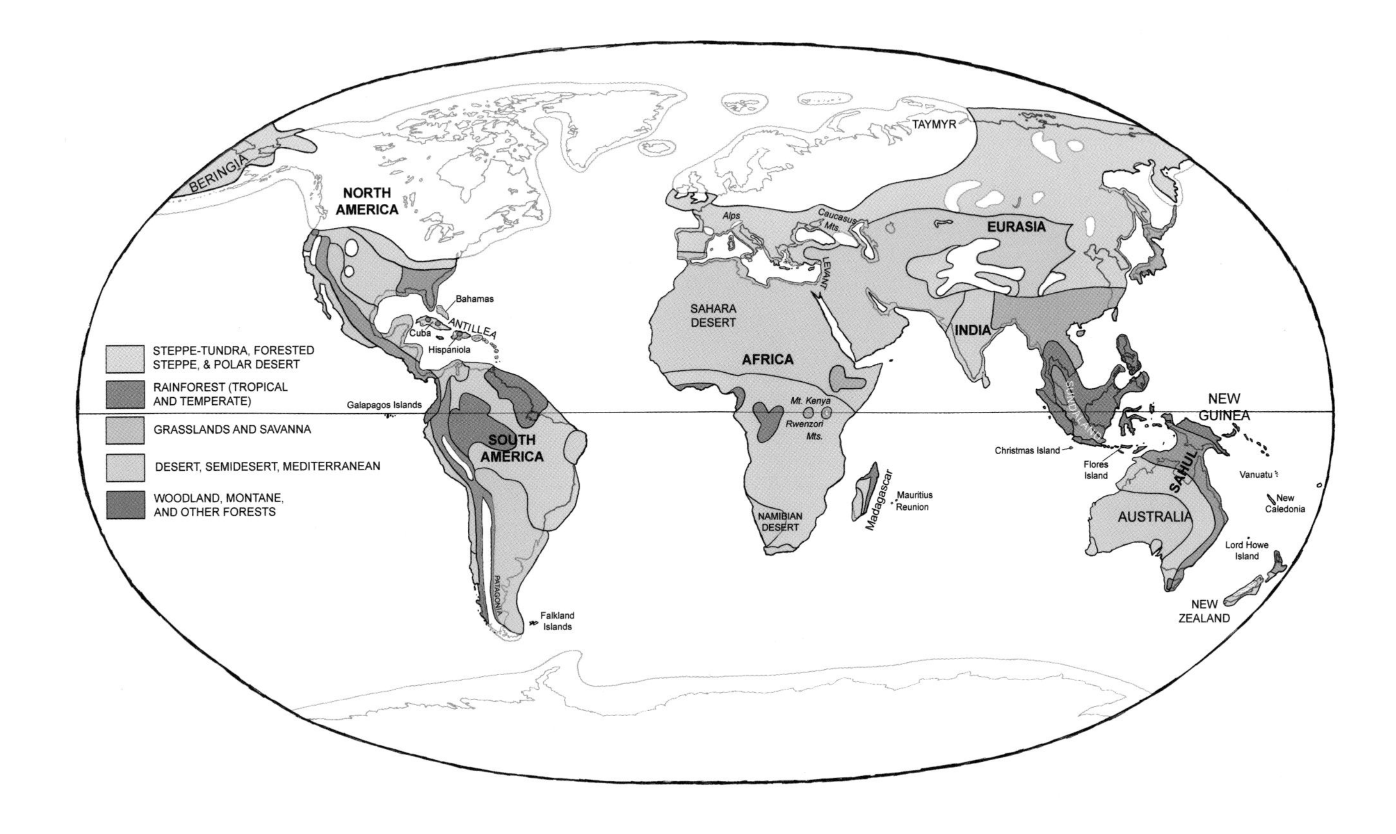

FIG. 3.5. **WORLD VEGETATION ZONES, 26,000 YEARS AGO:** The world as a whole was cooler and much drier in the Last Glacial Maximum. One indication of that was the great expansion of desert-like, grassland, and steppe environments. Humid tropical forests mostly retreated to small, discontinuous areas around the equator. Yet there were few vertebrate extinctions at this time.

East and western Asia grew enormously. Tropical forests were very restricted, with New Guinea and the islands of present-day Indonesia acting as important refugia (areas where species survival was still possible). In Australia conditions were extremely dry, with desert and grassland covering much of the continent.

In North America, the enormous area covered by glacial ice meant that the northern half of the continent, apart from ice-free central Alaska and Yukon, remained effectively uninhabitable during the Last Glacial Maximum. Farther south, with winters much colder than at present and cool summers, the eastern third of North America would have been clothed in coniferous boreal forest or taiga (see figure 3.5). The center would have been dry grassland, becoming increasingly desertic toward the west and southwest. Finally, dry woodlands would have occupied much of the west coast, with small pockets of temperate rain forest in the Pacific Northwest.

Northern Mexico at this time would have been clothed in scrub woodlands along both

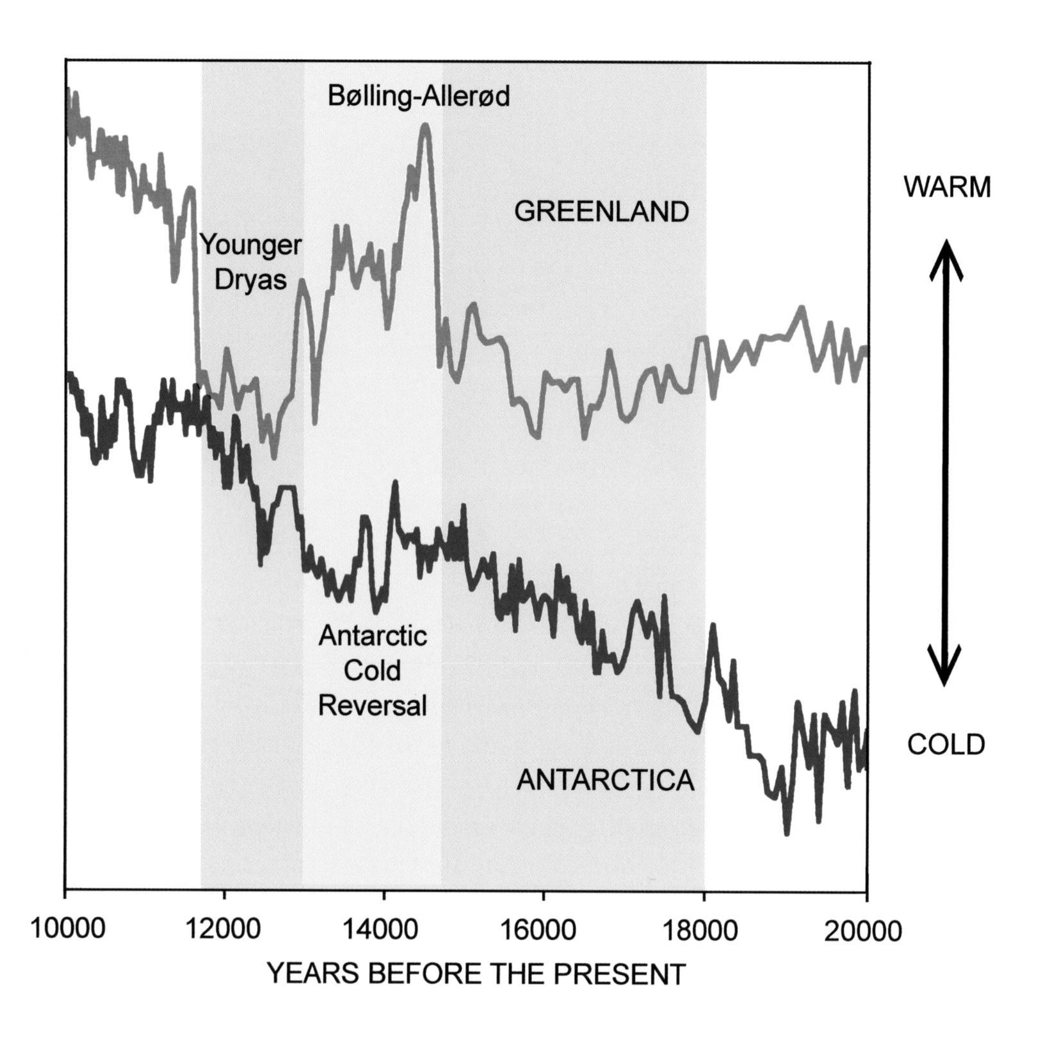

FIG. 3.6. **THE POLAR SEESAW:** The graph displays mean annual temperature records for Greenland (red line) and Antarctica (blue line) between 10,000 and 20,000 BP, as revealed by oxygen isotope data from ice cores. Although both curves record a general trend toward warmer temperatures across this time period, there are differences. When the Northern Hemisphere experienced notable warming beginning around 15,000 BP, the Southern Hemisphere became somewhat cooler. The reverse happened 12,900 years ago: temperatures plummeted to levels as cold as those of the Last Glacial Maximum in the Northern Hemisphere at the same time as the Southern Hemisphere was warming. The north started to sharply warm again after 11,700 years ago, whereas the south stayed more or less steady. Some authorities think that these sharp reversals in temperature trends, together with the initial arrival of people, forced extinctions at slightly different times in North and South America. The Younger Dryas qualifies as a Heinrich event (identified as H0), but the long-lasting Bølling-Allerød is usually considered an interstadial rather than a D-O event.

coasts, with very dry conditions in the middle part of the country. Southern Mexico, southern Florida, and all of Central America would have been dominated by dry tropical forest and grasslands, with humid forest only appearing close to the Panamanian isthmus. Similar conditions would have prevailed over the islands in the Caribbean (Antillea).

As the climate began to warm 18,000 years ago, major habitats would have moved northward in the wake of the retreating ice. As the Laurentide ice sheet retreated eastward toward Hudson Bay, grasslands and conifer-dominated boreal forest spread through Canada, bounded (as it is today) by tundra to the north and temperate deciduous forest in the east.

Ice Age conditions in South America were not the same as North America's because of its very different geography and general positioning within the equatorial region (see figure 3.5). The biggest single difference in vegetation today is the presence of tropical forest, which covers about one-third of the total area of South and Central America south of Mexico. South America does not possess a major latitudinal tundra like that of northern Canada, although there is altitudinal or alpine tundra along parts of the Andes, known as the páramo, and southern Patagonia is a dry steppe. Temperate deciduous forest, temperate rain forest, and conifer forest are mainly limited to slender strips along the southern Pacific margin.

At present, apart from local areas in the Andes, large-scale dry habitats—grasslands, savannas, and shrublands—mainly occur in two regions of South America: the Orinoco basin and related areas in the northwestern part of the continent, and a broad strip running from southern Patagonia to northeastern Brazil. These two zones of drier habitats are diagonally separated by tropical rain forest occupying the Amazon basin.

During the Last Glacial Maximum, lower temperatures and rainfall meant that the drier environments expanded considerably (see figure 3.5). Amazonia may have received 25 to 35 percent less precipitation than it does today, causing the rain forest to retreat into areas where rainfall remained high. Southernmost South America became very cold and dry, but conditions in midlatitude areas close to the Andes were better, as indicated by the many fossil sites of late date that have been discovered along both sides of the spine of this mountain chain.

Like the Rockies in North America, the Andes supported mountain glaciers for much of their length. A small ice sheet, the Patagonian, covered the southern Andes and extended out onto the exposed continental shelf (see figure 3.4). The Patagonian ice sheet's maximum area was probably on the order of 480,000 sq km (185,000 sq mi), amounting to less than 3 percent of the continent's total area 26,000 years ago (by comparison, ice sheets covered more than half of North America, up to 10 million sq km or 3.9 million sq mi).

Although Late Pleistocene conditions were very different from what we would regard as normal today, these fairly dry, grass-, steppe-, and woodland-dominated habitats clearly suited most of the really big herbivorous megafauna of the New World. In North America this group included bison, mammoths, and mastodons; in South America, camels, xenarthrans, meridiungulates, and gomphotheres (e.g., plates 8.3 and 9.5). Horses and camels

1. Shasta ground sloth *(Nothrotheriops shastensis)*
2. Horse *(Equus* sp.*)*
3. Western camel *(Camelops hesternus)*
4. Four-horned pronghorn *(Captomeryx furcifer)*
5. Harlan's ground sloth *(Paramylodon harlani)*
6. Stilt-legged llama *(Hemiauchenia macrocephala)*

PLATE 3.2. **SOUTHERN CALIFORNIA PANORAMA:** Rancho La Brea in Los Angeles is famous for its tar "pits," or natural seeps of asphalt, that acted as traps for unwary animals all through the Late Pleistocene. Every species in this plate is represented at La Brea by numerous individuals—hundreds of them in some cases. La Brea is best known for its large carnivores—sabertooth cat, false cheetah, short-faced bear, dire wolf, and so on (figure 1.3, and plates 6.1 and 11.4)—but Pleistocene southern California was also home to an extensive variety of herbivorous mammals. The camels, horses, and pronghorns represent native families that had evolved in North America over the course of millions of years; however, except for one surviving species of pronghorn, all had disappeared from the continent by the end of the Pleistocene. (Llamas managed to persist in South America, as did horses and other kinds of camels in Eurasia.) Recent studies indicate that the climate around La Brea at the end of the Pleistocene was not markedly different from what it is today—hot, dry summers with sparse winter rains, to which the mammals and birds were presumably well adapted.

1. Sable antelope *(Hippotragus niger)**
2. White rhino *(Ceratotherium simus)**
3. Giant Cape horse *(Equus capensis)*
4. Giant primal hartebeest *(Megalotragus priscus)*
5. Extinct savanna elephant *(Loxodonta atlantica)*
6. Plains zebra *(Equus quagga)** with quagga coloration

[* = an extant species]

PLATE 3.3. **SOUTHERN AFRICA PANORAMA:** The end-Pleistocene climate in the southernmost part of Africa would have been Mediterranean in type, featuring moderately wet winters and long, dry summers, much like today. There also would have been a diverse megafauna. The sable antelope and white rhino are both extant, but are now uncommon outside of reserves and game parks. The plains zebra is also an extant species, although the distinctive quagga coloration is no longer seen (plate 7.3). Although zebra were extensively hunted in the recent past because they were considered competitors of domestic horses and cattle, all species have survived. That's not the case for the giant primal hartebeest, represented by a female and her foal, and the elephant species seen on the upper slope, which was closely related to but larger than the living African savanna elephant. The big hartebeest persisted until at least 16,000 years ago, although it may have survived into the early Holocene according to some authors. It is known from cave sites in South Africa and was probably hunted. When the elephant *Loxodonta atlantica* finally disappeared is uncertain; it may have persisted into the Late Pleistocene, although most fossils are Middle Pleistocene in age.

lived in both places (see plates 3.2 and 9.1). Like the mammoth steppe in Eurasia, and for the same reason, these areas are notably rich in Quaternary fossils. By contrast, far fewer paleontological localities are known in Amazonia, partly because densely vegetated areas are difficult to prospect paleontologically, but also because truly large mammals were less likely to live there. Amazingly, despite the loss of the higher-latitude portions of North America to ice sheets and the desertification of Patagonia, the Last Glacial Maximum should have actually increased the potential range of habitats for species well adapted to drier conditions.

Africa did not experience an episode of concentrated megafaunal extinction during Near Time, but that simply makes it all the more interesting and important for our investigation. Very schematically, equatorial Africa is presently covered by tropical forest in the Congo basin and West African southern littoral. These areas are flanked by belts of grassland and woodlands. Farther away from the equator there are major midlatitude deserts, the Namibian in the south and the far larger Saharan in the north. These are in turn bordered by more mesic (moderately well-watered) zones along the Mediterranean coast and the south coast of southern Africa. All of these habitats underwent marked change and disruption during the late Quaternary, but except for the highest peaks, like Mount Kenya and the Rwenzori Mountains, Africa was not glaciated.

The enormous changes in the distribution of African tropical rain forest is a case in point. During the Last Glacial Maximum, previously continuous rain forest shrank into much smaller and discontinuous refugia, at the same time as nonforest vegetation, in the form of grasslands and sparse dry woodlands, expanded enormously (see figure 3.5). Climatologists think that Africa remained relatively cool and dry between 22,000 and 16,000 years ago, a period that in the Northern Hemisphere was a time of glacial retreat. If this scenario is substantially correct, then the large herbivores of Africa—elephants, horses, rhinos, and, especially, members of the cattle family Bovidae—would have been living in near optimum rangeland conditions throughout this interval (see plate 3.3).

By 15,000 years ago, Africa had become warmer and moister overall, which permitted the rain forest to reexpand. There was a new arid interval around 12,000 years ago during which the summer monsoon system dramatically weakened and lake levels fell in East Africa. This may have been correlated with the dramatic cooling interval in the Northern Hemisphere known as the Younger Dryas, which occurred at this time. Then, at the beginning of the Holocene, moister conditions returned in force. Increased rainfall enabled the appearance of pluvial lakes (i.e., lakes primarily fed by rainfall rather than by streams). The landscape was transformed into grassland and bush, thus becoming hospitable to large herbivores at the same time as their ecological counterparts were disappearing throughout the continental New World.

In the northern and central Sahara, there are several localities dating from 6000 to 8000 years ago that are decorated with rock art (pictographs and petroglyphs). The art clearly depicts large mammals and birds that are immediately recognizable as elephant, buffalo, rhino, hartebeest, lion, and ostrich, among others, all of which are still extant but no longer live in the region. Humans are frequently illustrated, bearing weapons, trapping apparatus, and apparent camouflage. Clearly, during the Pleistocene–Holocene transition and thereafter, large-bodied mammals were thriving in the well-watered grasslands of the Sahara, as were the local humans who both hunted them and captured them magically in their art. However, during the very same interval ground sloths, glyptodons, gomphotheres, and other large South American mammals disappeared, not only in dry Patagonia but also in Amazonia and everywhere else (see plates 2.4, 9.3, and 11.1). Little or nothing happened in Africa. The contrast is important.

Although this introduction to Ice Age climate and vegetation is necessarily brief, I want to mention two other kinds of climatological events that are still poorly understood but might be important for understanding the cause of some extinctions. These are the short-term, irregularly spaced episodes known as Heinrich (cooling) and Dansgaard-Oeschger or D-O (warming) events, in which mean annual temperatures swung as much as ±7–8°C (12.6–14.4°F) over short periods (decades to centuries) at high latitudes. What causes these upticks and downticks is debated, but there is evidence for both kinds of events in the Late Pleistocene, some occurring very close to the Pleistocene–Holocene transition. Although these temperature excursions certainly look harsh on paper, it is hard to know what their real-world effects might have been. For one thing, in temperate and equatorial regions the effects of dramatic temperature change in polar regions would have been damped by atmospheric mixing, which presumably would have reduced their biological impacts. Furthermore, nothing of similar magnitude has occurred during the past 10,000 years, so Holocene humans haven't (yet) had any experience with long-term consequences of short-term severe cooling or warming. The so-called Little Ice Age (conventionally, 1300–1850 CE), which was not really an ice age and perhaps not even a synchronous event globally, witnessed a mean annual temperature drop of just 1–2°C (1.8–3.6°F) in parts of the Northern Hemisphere. Yet it was enough to freeze the lower Hudson River in the winter of 1780, famously allowing New Yorkers so inclined to walk from Manhattan to Staten Island. And as far as we know, no vertebrate extinctions can be linked to the Little Ice Age. However, the point remains that prehistory shows that climate change can happen extraordinarily quickly. We may not be so lucky the next time. . . .

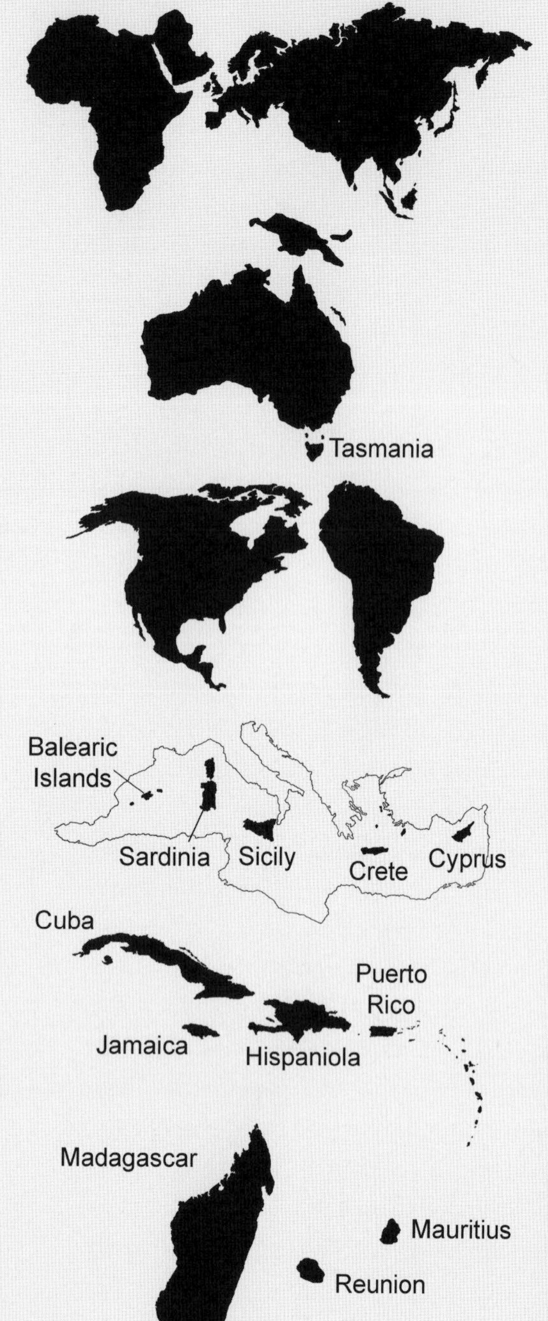

Africa & Eurasia: No major, concentrated episode of extinction during past 100,000 years. Losses uncorrelated with biological first contact (anatomically modern humans present in Africa by at least 350,000 BP).

Sahul: Major extinction interval 40–45,000 BP, long after human arrival 65,000 BP. Losses occurred during period of significant transformation of habitats.

Continental Americas: Humans arrived at least 15–16,000 BP, but somewhat earlier arrival cannot be excluded. Major extinction episode terminates around 11,000 BP; few extinctions thereafter.

Mediterranea: Humans traveled to larger islands no later than 10,000 years ago, but much earlier presence possible in some cases (e.g., Sardinia). Extinctions poorly dated but extended over thousands of years, long after first contact.

Antillea: Humans present by 6,000 BP. Sloth extinctions begin 4,000–5,000 BP, long after first contact.

Madagascar: Archeological evidence for human presence no earlier than 4,000 BP. Extinctions possibly delayed until 2,000 BP, with final premodern losses occurring as late as 1600 CE.
Mauritius & Reunion: Uninhabited prior to European discovery in 16th century CE. Extinctions within decades of settlement and importation of rats, domesticated animals.

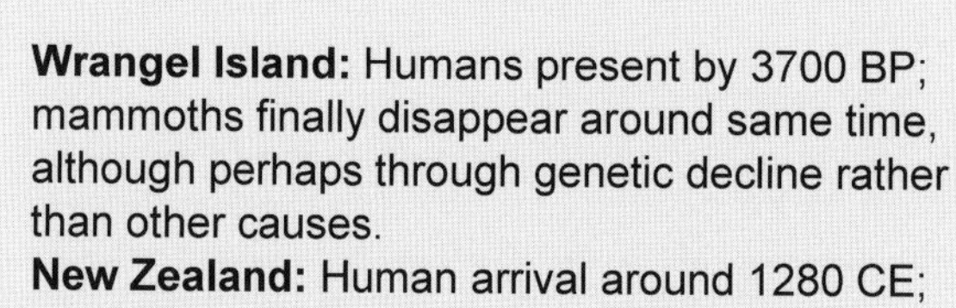

Wrangel Island: Humans present by 3700 BP; mammoths finally disappear around same time, although perhaps through genetic decline rather than other causes.
New Zealand: Human arrival around 1280 CE; major extinction episode within one century.

4

THE HOMININ DIASPORA

FIG. 4.1. **THE HOMININ DIASPORA AND BIOLOGICAL FIRST CONTACT:** There is no single pattern of loss during Near Time. Extinction was fast on some islands, slower on others. Continental extinctions were strongly concentrated in time in the Americas but not in Eurasia nor in Africa, which lost the fewest megafauna.

Although it is difficult to believe in light of the current size of the world's human population (7.6 billion and counting), there was a time when we were not absolutely everywhere on the planet. How we got to be everywhere begins with our early diaspora, or dispersal, with its eventual fateful impacts on floras and faunas the world over (see figure 4.1). Although I will be concentrating on the spread of our own species, we need to begin with the stage-setting journeys undertaken by some of our close ancestors and relatives—other species in the genus *Homo*, collectively known as hominins.

AFRICA, EURASIA, AND SAHUL

As far as we can now tell, our evolutionary trail started in Africa about 2 million years ago, with the appearance of *Homo heidelbergensis*, the earliest broadly accepted member of our particular portion of the hominin family tree (see plate 4.1). When this species first spread to Eurasia is quite uncertain, but in round figures it could not have been much later than about 1.5 million years ago, because there is evidence for the presence of hominins much like *H. heidelbergensis* in far-off Java around that time. If these Javanese hominins had watercraft, as has been conjectured, in principle they should have been able to keep going all the way to present-day Australia and New Guinea. At present, there is no evidence of Middle Pleistocene hominins in either portion of Sahul, so we must assume that they didn't make it that far. *Homo heidelbergensis* also got to Europe (where the first fossils belonging to this species were found, hence its non-African name). One widely held view is that the European population gave rise early in the Pleistocene to *Homo neanderthalensis* (Neandertals) and possibly the Denisovans, a genomically and presumably morphologically distinct group of humans that has not yet been given a formal name (see plate 4.2).

At the time *neanderthalensis* was emerging in Europe, the evolution of stay-at-home *heidelbergensis* continued in Africa, eventually leading to the appearance of our species, *Homo sapiens*, on that continent. With our very large brains, small jaws, and gracile skeletons, we can be distinguished, as "anatomically modern humans," from our more archaic (but still human) antecedents. However, the dividing line is not sharp. Anatomically

PLATE 4.1. **EARLY HOMININ BUTCHERING A WARTHOG** in southern Africa. In this plate a member of the species *Homo heidelbergensis* is seen skinning an example of the extinct warthog *Metridiochoerus modestus*. Interestingly, pig remains are common in African hominin sites of many different ages, indicating that suids were an important source of protein. The extant warthog (*Phacochoerus africanus*), by definition a megafaunal species (50–150 kg/110–330 lb), has been heavily persecuted in modern times, but it is still found throughout Africa south of the Sahara, in both forest and savanna.

PLATE 4.2. **A NEANDERTAL HUNTER BRINGS HOME A ROE DEER** (*Capreolus capreolus*). Neandertals and anatomically modern humans shared a close common ancestry and were able to hybridize to some degree, as ancient DNA studies have shown. Hominins with Neandertal features—including stocky build, heavy brow ridges, and remarkably large skulls—appear in the fossil record about 400,000 years ago. Then they disappear, rather mysteriously, about 40,000 years ago. Although it is clear from their tool kits and occupation sites that Neandertals hunted megafauna (plate 9.4), there is no evidence that they caused any outright extinctions. Nor is there any evidence that *Homo sapiens* caused the extinction of Neandertals, with whom they significantly overlapped geographically for tens of millennia.

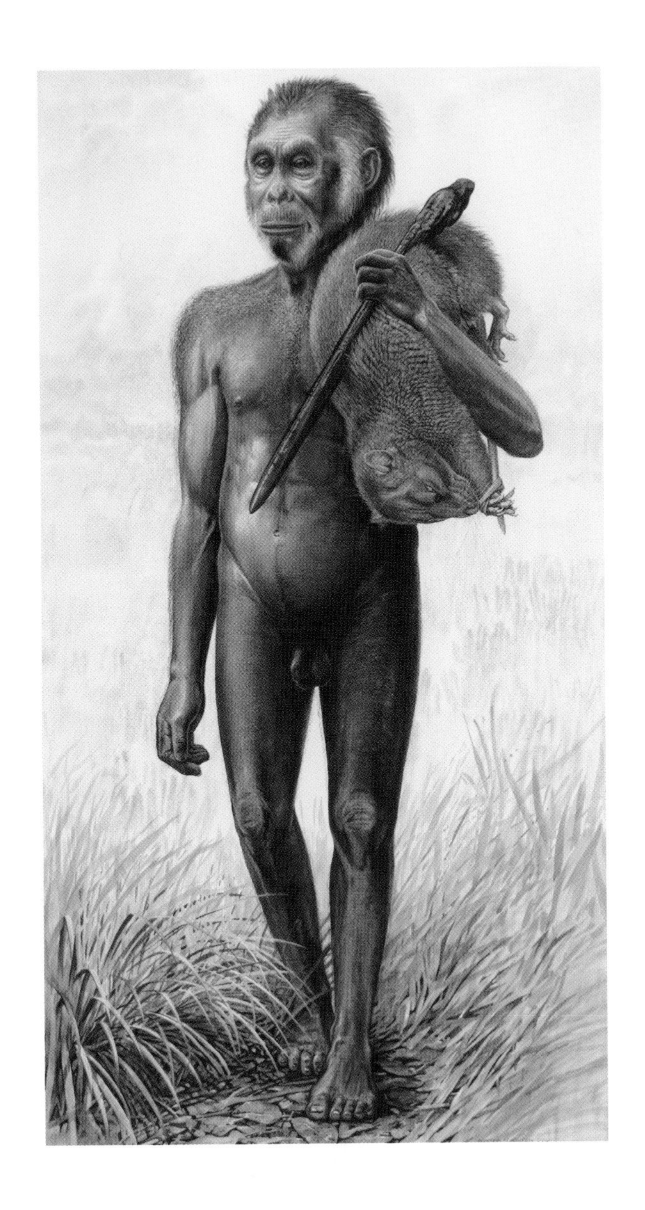

PLATE 4.3. **THE RELATIONSHIPS OF THE "HOBBIT"** *(Homo floresiensis)* to other hominin species have been hotly contested since the initial discovery of "Hobbit" remains in 2003, at the site of Liang Bua on Flores Island, Wallacea (figure 3.4). Their diminutive size (1.1 m/3.5 ft) suggested to some commentators that they were just small-statured, inbred modern humans who had suffered from genetic or developmental diseases. This idea has been discredited, but debate continues regarding whether *H. floresiensis* was an advanced hominin related to ourselves, but whose lineage underwent size reduction as a possible adaptation to island life, or was instead derived from a more primitive ancestor, such as small-brained *H. erectus*. Whatever its evolutionary status, archeologists have demonstrated that *H. floresiensis* had fire, made stone tools, and hunted local animals such as large turtles, the pygmy elephant *Stegodon florensis*, and, as seen here, the giant rat of Flores, *Papagomys theodorverhoeveni*. This hominin disappears near the beginning of Near Time, just after *H. sapiens* shows up in the Indonesian fossil record.

modern people were already living at the Moroccan site of Jebel Irhoud roughly 350,000 years ago (see figure 4.1), an appearance date which is at least 100,000 years earlier than estimates made only a few years ago on the basis of the human fossil record as it was then known. This surely won't be the last word on when and where our immediate ancestors arrived on the scene.

From about 120,000 years onward, advanced hominins were evidently consistently present in what is now insular southeast Asia in the form of at least two distinct lineages, namely our own *sapiens* branch and the small-sized species playfully called the "Hobbit," *Homo floresiensis* (see plate 4.3). From there, *H. sapiens* finally made the crossing to Sahul no later than 45,000 years ago, and perhaps as early as 65,000 years ago, presumably via Sundaland (the platform, now partially submerged, on which the islands of central and western Indonesia are situated; see chapter 8). As noted, part of this journey had to have been made by boat, because even during times of sea-level lows Sahul was never connected to the islands comprising western Indonesia (see figure 3.4). The antiquity of human occupation of this part of the world is not nearly settled, in part because physical remains of these earliest sojurners have yet to be found (see plate 4.4).

In light of the main theme of this book, it is important to mention there were no identifiable periods of concentrated megafaunal extinctions in the Old World during the expansion of premodern or non-*sapiens* humans. There were losses now and again, but they occurred without any apparent relationship to one another or to the recency of premodern hominin arrival in a specific area (see figure 4.1).

CONTINENTAL NEW WORLD

The evidence for early human presence in the New World is scanty and disputed, and there are a number of currently competing claims regarding the time of initial entry into North America. Over the past two decades the admittedly conservative archeological consensus date for first entry has been incrementally pushed back, bit by bit, from roughly 12,000 years ago when Paul Martin was first writing, to around 15,000 years ago at present. Pushing hard against this consensus are insurgent claims for much earlier human presence in the New World. We will focus on recent work and its implications for the extinction debate (see figure 4.2 for site locations).

Prior to the late 1960s, it was generally believed that humans could have gained entry into the midcontinent only by traveling out of Beringia along the eastern slope of the Rockies. There, they could have used the so-called ice-free corridor, or the gap between the facing margins of the Cordilleran and Laurentide ice sheets (see figure 3.3). The corridor permitted faunal movements in both directions, and among other things it would have permitted different populations of the same species to keep in genetic contact across northern North America, at least in principle. When the ice margins coalesced, however, such movements

PLATE 4.4. **A GIANT SNAKE** (*Wonambi naracoortensis*) and a human observe each other, carefully, in southern Australia 65,000 years ago. The snake's genus name is derived from an aboriginal word and refers to an enormous serpent from the Dreamtime (time immemorial, populated by ancestral or supernatural beings). Some authors have suggested that the wonambi was actually seen by Australia's early inhabitants, who passed down stories about it across generations. Although not a python, the wonambi was a constrictor, and would have killed its prey in a similar manner. Fairly complete specimens of this huge snake have been recovered from the Naracoorte caves (plate 7.4), indicating that it attained a length of 5–6 m (16–19 ft) and a body size of around 100 kg (200 lb). Some animals may have been significantly larger.

would have been out of the question. This last happened during the Last Glacial Maximum, when the ice sheets attained their largest dimensions. Over the years there has been much discussion regarding when, and for how long, the corridor would have been remained closed. Recent studies have fixed the last period of closure as the interval between 23,000 and 13,400 years ago, which means that first entry, at least under the corridor theory, had to have taken place either before or after those dates. The latter possibility can now be discarded, thanks to the discovery of several archeological sites that are as old as or older than 13,400 BP. Of these, the most significant for our purpose is the Page-Ladson mastodon kill site in the panhandle region of Florida (see figure 4.2), which has yielded both unquestionable artifacts and radiocarbon dates as old as 14,500 BP. Thus, if people first reached the

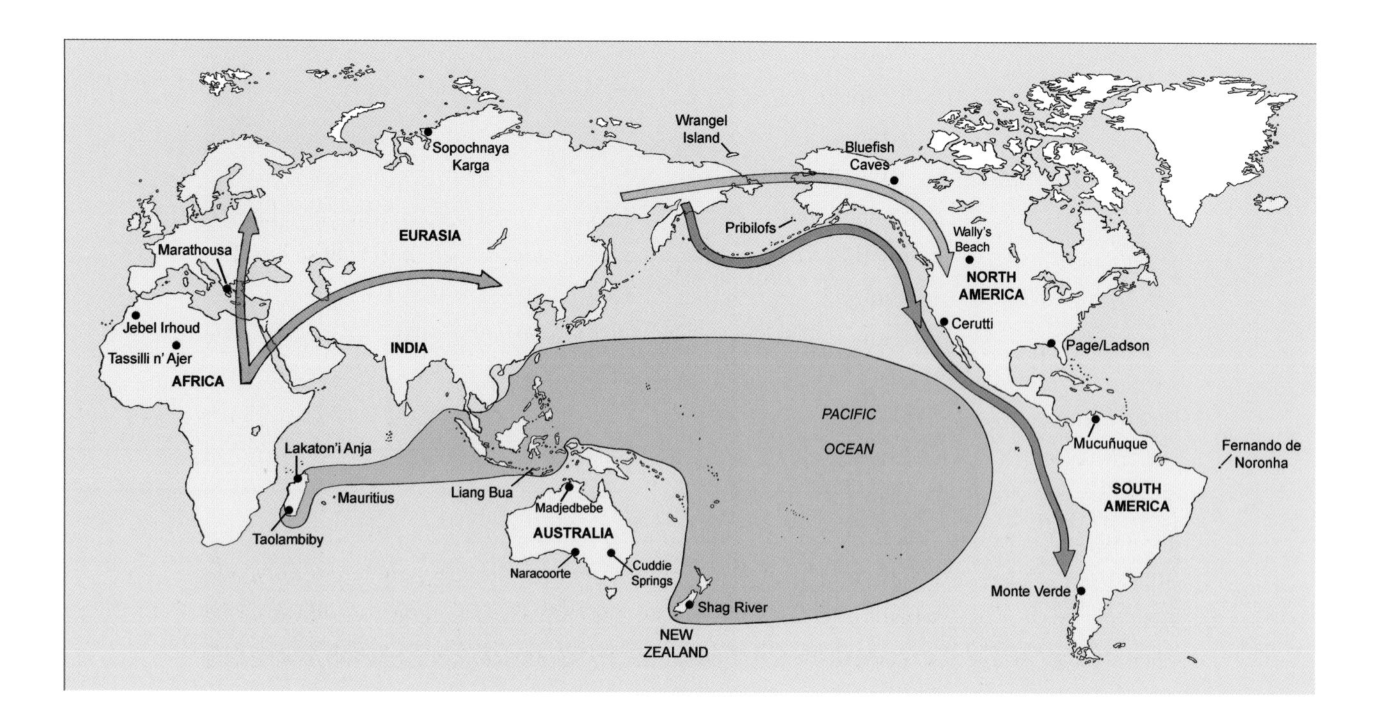

FIG. 4.2. **OUT OF AFRICA, AND ALSO OUT OF ASIA:** Modern *Homo sapiens* arose in Africa more than 350,000 years ago and entered Eurasia thereafter (red arrows), probably on several occasions. People were in northernmost Siberia by at least 45,000 years ago, and may have begun migrating across the Bering land bridge into North America as early as the Last Glacial Maximum, 23,000 to 27,000 years ago. However, there is no conclusive evidence for their presence in midcontinental areas until about 15,000 years ago, which they might have attained by passing along the ice-free corridor before its closure, or by coasting along the Pacific seaboard. Or perhaps they did both (orange and blue arrows). The Austronesian diaspora (green area) occurred much later, starting in the middle Holocene. Leaving their heartland in southeast Asia, Austronesians eventually spread out over vast distances, as far as Easter Island in the Pacific Ocean and Madagascar in the Indian Ocean. Their migrations culminated in the thirteenth century CE with arrival in New Zealand and other far-flung islands in the western Pacific. (This map also identifies important paleontological localities discussed in the text.)

heartland of North America by the corridor route along the Rockies, they would have had to have been on the move no later than 23,000 years ago. Can this be backed up?

There are a number of archeological sites for which a claim of Last Glacial Maximum age has been made, but the one with the best credentials at present is the Bluefish Caves locality in northern Yukon (see figure 4.2). Although stone artifacts have not been recovered from levels older than about 13,000 years ago, bones bearing apparent cut marks have. The oldest of these bones has recently returned a radiocarbon date of 24,000 BP. Assuming that the cut marks are correctly interpreted, the Bluefish material indicates that humans were indeed present in the northern part of North America at least as early as the middle of the Last Glacial Maximum. This, however, is not evidence that they were also *south* of the ice at this time. The oldest Bluefish dates lie perilously close to the time that the corridor was being closed off, suggesting that perhaps the first migrants from Asia didn't get so far after all, but instead remained bottled up in eastern Beringia until a much later date.

There is, however, an alternative theory regarding initial human entry that does not involve the ice-free corridor at all. The idea here is that people could have passed down the west coast of North America by boat, or traveled by foot along then exposed parts of the continental shelf lining the Pacific rim, at any time during the Late Pleistocene. In this scenario (see figure 4.2), the interior of North America could have been penetrated by simply skirting the southern margins of the ice sheets, making entry into the midcontinent somewhere between Washington and California. Whenever this happened, if it did at all, the date on the Page-Ladson site practically demands that either people walked very, very quickly to get to Florida from the Pacific coast, or they had already been spreading through North America for quite some time.

There are other theories, among which is the provocative argument made by Bruce Bradley of the University of Exeter and Dennis Stanford of the Smithsonian Institution that perhaps the entry point into North America was not in the west but in the east, somewhere in the Maritimes or New England. The theory in this case is that the first migrants actually came from western Europe, and undertook their journey by traveling by boat along the sea-ice margin. However, the hard evidence for this consists of certain formal similarities in how stone tools were made by Solutrean and Clovis peoples on opposite sides of the North Atlantic. The similarities that Bradley and Stanford have described are noteworthy, but are they evidence of cultural contact or even migration? The general opinion among archeologists is that the resemblances are simply due to technological convergence—that is, to the separate invention of similar techniques, in this case specific ways of working in stone. Moreover, this hypothesis has come up against a strong counterargument that seems to leave little room for doubt about the identity of the first humans in the New World. Paleogeneticist Bastien Llamas and coworkers recently compiled genetic evidence from several dozen pre-Columbian skeletons from South America in order to determine their degree of relatedness and the likeliest date for their ancestors' entry into the New World. According to their calculations, those ancestors arrived around 16,000 years ago—and were clearly derived from populations living in eastern Siberia, not western Europe. This finding does not completely exclude the possibility that early Europeans separately entered eastern North America, but in the absence of any equivalent supporting evidence, the Solutrean hypothesis is best regarded as doubtful.

We shall now consider the even more challenging hypothesis that humans were

PLATE 4.5. **MASTODONS IN MUSTH?:** Although American mastodons (*Mammut americanum*) look characteristically elephantlike, they retained a number of ancient proboscidean features, particularly evident in their teeth. Their diet probably varied over the course of the year, as well as regionally, but fossil stomach contents indicate that mastodons were mostly browsers. They also had a wide range: during favorable times mastodons lived from Central America to Yukon and Alaska. This scene, of two males engaging in combat, is actually based on real discoveries involving broken ribs, evidence of goring, and tusks locked in a death grip. Male elephants today become highly aggressive during what is known as musth, when their testosterone levels skyrocket and animals fight over dominance positions. Mastodons may have been similar.

in the New World *much* earlier, but have eluded detection until now. Is this even remotely likely? Steve Holen and Tom Deméré of the San Diego Natural History Museum and their colleagues certainly think so, because they recently made the startling claim that a mastodon skeleton, found at the Cerutti site in southern California and dated to around 130,000 years ago, bore evidence of butchering, or at least preparation activities associated with breaking long bones to get at the contained marrow. Apparent artifacts, in the form of rocks evidently transported from elsewhere, have been identified at the site, but there's no physical evidence of the people themselves who might have been at work on the carcass.

Well-preserved mastodon remains are not uncommon in North America, especially in the Northeast and Great Lakes regions (plate 4.5), and there are several unquestionable examples of mastodon kill sites. But given that the Cerutti mastodon's date is almost ten times older than the fragile consensus date for the first presence of hominins in central North America, resistance from archeologists has been keen. There has also been some collateral speculation that the Cerutti hominins may have been Neandertals or some other archaic type of human rather than thoroughly modern people, but this just trades one imponderable for another. (The oldest dated human remains from anywhere in the New World, from a submerged cave site in Yucatan, are only about 12,000 years old.) Something just seems wrong with the Cerutti find: either the dating is way off (unlikely) or the interpretation of the alleged artifacts and pattern of bone breakage is in error.

Another question of interest is whether some of the earliest human travelers to the New World kept on going southward, continuing all the way to the tip of South America (see figure 4.2). The site of Monte Verde in southern Chile, which has yielded evidence of human occupation in the form of tools, hearths, and even the remains of a possibly butchered gomphothere, is seriously old. The earliest phase of human occupation at Monte Verde was originally placed at 14,500 years ago. This was controversial enough at the time of the first reports on the site in the 1980s, but the initial occupation date has now been pushed back to 18,500 years ago on the basis of more recent age estimates using radiocarbon and OSL (optically stimulated luminescence; see "Appendix: Dating Near Time," page 191). If this older age is correct, it is staggering in its implications: it means that humans could have been present in the southern part of South America before they were in midcontinental North America.

Reconstructions of the local landscapes around Monte Verde at this earlier time indicate that the locality would have been situated close to glaciers in the southern Andes, still at their near maximum extent 18,000 years ago. This means that the humans would have been occupying an environment with strong seasonal variation and prolonged cold conditions in winter—a hard place to live, although perhaps not that much different from Beringia. Apart from specialist disagreements over the nature of the occupation itself, a real difficulty with accepting Monte Verde's dating at face value is that, like the Cerutti site in California, there is no other accepted site of equivalent age anywhere along the coasts of North or South America. Until additional, credible evidence of early human occupations resembling Monte Verde has been found, its significance will have to remain uncertain.

1. Dwarf Maltese hippo *(Hippopotamus melitensis)*
2. Dwarf Maltese elephant *(Palaeoloxodon falconeri)*
3. Giant Maltese swan *(Cygnus falconeri)*

PLATE 4.6. **MALTA PANORAMA:** Many Mediterranean islands supported native species of mammals and birds that have now completely disappeared. Elephants, deer, and hippos were frequently present in these faunas, bearing witness to their ability to colonize islands. As elsewhere, large mammals tended to undergo strong selection for downsizing: the dwarf Maltese elephant illustrated here was no larger than a small pony; the tiny hippo was the size of a large pig. By contrast, island-adapted birds often grew much larger than their mainland ancestors, in part because they no longer needed to retain small body sizes in order to fly efficiently. The giant Maltese swan, for example, was probably about 25 percent larger than the North American trumpeter swan (*Cygnus buccinator*), the largest living waterfowl. Although the dwarf elephant and hippo lasted until the end of the Pleistocene, the swan is thought to have died out much earlier.

THE ISLANDS

First human arrival is very poorly documented for most islands, and for this reason evidence of faunal collapse, if any, is often used as proxy evidence for human presence. This is dangerous for extinction studies because it assumes what is to be proven, and I only briefly touch on such evidence in this section (see chapter 8). Locations of islands mentioned in this and other chapters can be found in figures 4.1 and 4.2.

Mediterranea (see plate 4.6) is a collective name for the many small to medium sized islands that dot the Mediterranean Sea. However, this grouping is not meant to imply that they all have the same geological or extinction history. Mediterranea includes the major islands of Cyprus, Crete, Sicily, Corsica, Sardinia, and the Balearics, as well as many small islands in the Aegean Sea, many of which were inhabited by species now extinct. Although some of the big islands would have been moderately larger during times of lowered sea level, with the exception of Sicily none was continuous with surrounding mainlands at any time during the Pleistocene. Thus the earliest human occupants could have only reached these islands over water, island by island, perhaps over a lengthy time span.

Premodern humans like *H. heidelbergensis* were in southern Europe no later than about 750,000 years ago, because sites of this age are known from Spain and elsewhere. Although claims for the presence of species other than *Homo sapiens* in Sardinia and Crete have been made in the past, the evidence is thin. On most islands, the earliest well-analyzed cultural levels are identifiably Neolithic, which suggests that Mediterranea was not extensively settled until the beginning of the Holocene.

Alan Simmons of the University of Nevada has argued that early settlers were hunting dwarf hippos and perhaps other species on the south coast of Cyprus at the end of the Pleistocene. His evidence comes mainly from a cliffside cave known as Akrotiri Aetokremnos, which contains a great number of dwarf hippo bones, some burnt, as well as quantities of stone flakes and other artifacts. Although it is hard to imagine how (or why) the hippos would have congregated in the cave unless human intervention were involved, not a single bone displays a cut mark. Similar arguments have been made for fossils found in cave sites elsewhere in Mediterranea, but confirmatory evidence for human hunting is essentially nonexistent.

Antillea (see plate 4.7) includes all island landmasses within the Caribbean Sea, excluding shelf islands like Trinidad. A number of extinctions occurred in Antillea during the mid-Holocene and later but most are poorly dated. Even so, it is clear that these islands sharply contrast with the continental parts of the New World in one important regard: the earliest dates of human occupation. Limited archeological evidence implies that early settlers reached Cuba, Hispaniola, and Puerto Rico no earlier than roughly 6000 years ago, which is at least 8000 to 9000 years later than the earliest widely accepted occupancy date for the continents. Jamaica, whose archeological record is especially slim, may not have

PLATE 4.7. **FOREST SCENE IN CUBA:** The sheep-sized *Megalocnus rodens*, seen here confronting a Cuban crocodile (*Crocodylus rhombifer*) at the famous paleontological site of Ciego Montero, was the largest of the half dozen species of sloths that inhabited Quaternary Cuba. Although probably largely ground-dwelling, *Megalocnus* possessed a full set of large claws and powerfully built limbs that could have equally served for digging or limited climbing. Its species name ("gnawing") is a reference to its front teeth, which were greatly enlarged and ever growing, something like those of a rodent. Why such a specialization appeared is uncertain, but it suggests that this species was adapted to eating tough foods, such as tubers or fruits with husks. The Cuban crocodile is extant but critically endangered. It prefers freshwater environments and is notably terrestrial. In paleontological sites, remains of extinct mammals and birds have occasionally been found with distinctive round puncture marks suggestive of crocodile bites. *Megalocnus* was still extant as recently as 4000 years ago, although claims of association between sloth remains and cultural evidence have never been substantiated. The extinct Cuban crane (*Grus cubensis*), also seen here, had reduced wings and was flightless.

1. Malagasy riverine hippopotamus *(Hippopotamus lemerlei)*
2. Malagasy pygmy hippopotamus *(Hexaprotodon madagascariensis)*
3. Madagascar crocodile *(Voay robustus)*
4. Indian darter *(Anhinga melanogaster vulsini)**
5. Giant Malagasy tortoise *(Geochelone grandidieri)*
6. Muller's large elephant bird *(Mullerornis grandis)*
7. Baboon lemur *(Archaeolemur edwardsi)*
8. Giant sloth lemur *(Archaeoindris fontoynonti)*

[* = an extant species]

PLATE 4.8. **WESTERN MADAGASCAR HIGHLANDS PANORAMA:** Western forests in Madagascar are fairly open, with huge baobabs looming over other trees like solitary giants. In the past, gallery forests growing along river margins, as shown here, would have provided habitat and protection for many kinds of vertebrate species. Both the Malagasy riverine hippopotamus and the Malagasy pygmy hippopotamus were small compared to the living East African hippo. The pygmy hippo may have been quite terrestrial and is thus depicted grazing along the riverbank, rather than basking in the water or on the shore like its riverine cousin. The extant Nile crocodile was seemingly everywhere in subfossil times and still occurs today in northern Madagascar. However, depicted here is its native relative, the Madagascar crocodile, which disappeared along with the rest of the subfossil fauna less than a thousand years ago. A family group of semiterrestrial baboon lemurs skirts the riverbank, on its way to feed in the forest. Overhead, a giant sloth lemur cautiously hangs, like an overripe fruit. Farther back from the river we see a giant tortoise and a group of Muller's large elephant birds. An Indian darter, the cormorantlike bird spreading its wings on the back of an obliging hippo, is the only extant animal in this scene.

been occupied until 1500 years ago. The Lesser Antilles appear to have been occupied relatively late as well, despite the proximity of the chain to northern South America.

Madagascar (see plates 4.8 and 4.9), the world's fourth largest island, lies just 450 km (280 miles) east of the African coast, but, like Antillea, it was not occupied by people until very late. Until recently it was generally believed that initial settlement, by people of Austronesian origin, took place only about 2000 years ago. The so-called subfossil fauna of megafaunal lemurs, giant birds and other unique vertebrates seems to have mostly disappeared by about 1000 CE, although a few (including at least one kind of native hippo) persisted until shortly before (or even somewhat after) European discovery around 1500 CE.

The accepted chronology of human arrival was recently thrown into question by the late Bob Dewar of Cambridge University, who, with colleagues, discovered evidence of human occupation on the northern side of Madagascar going back nearly 4000 years, or twice the length of time previously accepted. The evidence is unprepossessing—as the scientists put it, it consists of only "a small sample of tiny items," mostly stone flakes, found at the site of Lakaton'i Anja and one other. Their preliminary study of this material failed to disclose any obvious link to the material culture of later inhabitants of the island. What had happened to these first inhabitants? Were they transients? Did they fail to settle the island successfully? The record is silent.

What we do know, however, is that humans were already interacting with the subfossil fauna by 2000 BP, because butchered hippo bones of this age occur at the site of Taolambiby in the southwestern part of the island. In recent years, additional examples of modified remains have been discovered, although the quantity is still small. None are as old as the oldest archeological sites from 4000 years ago, although paleontological sites of this age (and older), with no evidence of human presence, are now known from many parts of the island. Although Martin's case for extremely fast extinction in Madagascar is not refuted by the new age estimate for first human arrival, it is weakened.

New Zealand (see plate 4.10) consists of two very large islands (South and North Islands, respectively the twelfth and fourteenth largest in the world in overall size) and a flock of smaller ones lying close by. The big islands were the last large habitable landmasses to be discovered and settled by humans, an event that occurred around 1280 CE according to excellent radiocarbon records. The original Maori occupants were Austronesian by ancestry, just like the first settlers of Madagascar. They may have comprised a group of very modest size, perhaps consisting of no more than several dozen individuals according to some recent estimates. More than thirty bird species, including all the large flightless species known as moa, became extinct within a century or so of human settlement. No other islands with a large native fauna suffered as many losses in so short a period of time. New Zealand is unique in another sense: no other islands have yielded as much evidence of human hunting, in the form of worked or otherwise modified bones of extinct species, kill sites containing large numbers of individuals, and occupation sites with food refuse of butchered animals.

PLATE 4.9. **THE BIBYMALAGASY** ("animal of Madagascar," *Plesiorycteropus madagascariensis*) was surely one of the stranger elements in the island's subfossil fauna. The size of a small dog, nonmegafaunal *Plesiorycteropus* apparently lacked teeth and may, like an aardvark or an anteater, have fed mainly on social insects. Its external appearance is unknown; a remote possibility, suggested here, is that it may have been covered in scales like the living (and quite unrelated) Afroasian pangolins. The bibymalagasy's distinctive fossils have been discovered in several parts of the island, but it never shows up in human occupation sites and was probably not hunted. Nor is there any evidence bearing on its extinction date, although this presumably occurred before 1500 CE.

Some amount of forest clearance might have contributed to the endangerment of the megafaunal moa when they were still extant, but this likely applies only to lowland areas. In short, New Zealand presents exactly the combination of pattern features one would expect to see wherever prehistoric overhunting was supposedly conducted, but almost never does.

Other habitable islands occur in profusion in tropical and temperate regions—far too many for summary coverage here. Many of the ones that supported now extinct endemic vertebrates lie in the South Pacific and, like New Zealand, were colonized by humans only within the last millennium. Others lie far from any mainland in the Indian and Atlantic

1. South Island giant moa *(Dinornis robustus)*
2. Elephant-foot or thick-set moa *(Pachyornis elephantopus)*
3. Bush moa *(Anomalopteryx didiformis)*

PLATE 4.10. **CANTERBURY PLAINS PANORAMA:** The Canterbury Plains on the eastern side of South Island is one of the most productive fossil areas in New Zealand. At center stage in this depiction are two giant moa dancing in a courtship ritual—the larger one being the female. Some moa showed remarkable levels of sexual dimorphism, with females weighing as much as 200 kg (440 lb), or 2.5 times the size of their males. Until this situation was revealed by studies of ancient DNA, taxonomists usually placed males and females in different species. There were probably between nine and twelve species of moa living in late Holocene New Zealand, before the arrival of people in the thirteenth century CE. Their bones have been found in many different contexts on both main islands, from montane regions to coastal plains, from humid forest to dry grasslands. This strongly implies that moa were ecologically highly diversified, but that did not save them: all were probably gone within a century of human arrival. Richard Owen, the nineteenth-century anatomist who first brought scientific attention to moa, remarked that "such bulky and probably stupid birds, at first without the instinct and always without adequate means of escape and defence, would soon fall prey to the progenitors of the present Maoris."

Oceans and were never inhabited or even seen by people before 1500 CE. Many of them supported species found nowhere else, and almost all of these experienced losses as soon as humans showed up, early or late.

Two examples will have to suffice. Fernando de Noronha (see figure 4.2), a small group of Atlantic islands near northeastern Brazil, was uninhabited when first visited in 1503 by an expedition that may have included the famous cartographer-explorer Amerigo Vespucci. A "large rat" was seen on the main island in the group, but never recorded again. Researchers Michael Carleton and Storrs Olson of the United States National Museum of Natural History, who named the species *Noronhomys vespuccii* on the basis of fossil evidence recovered from the island, concluded that it must have died out soon after European discovery, perhaps as a result of competition with or predation by black rats (not closely related to *Noronhomys*, despite all of them being called "rats").

The island of Mauritius near Madagascar (see figure 4.2) was likewise uninhabited when first sighted by Portuguese vessels, around 1513 CE. It has the distinction of being the only place in the world where the dodo (*Raphus cucullatus*) lived. Settlers came in around 1600; the last credible sighting of the dodo occurred in 1662. However, its loss was probably not due to overhunting. As one wag memorably put it, for humans the stringy meat of the dodo would have been about as tasty as a tennis racket, and the bird was probably mostly ignored by the residents of Mauritius. Not so its eggs, which would have been of great interest to introduced rats, cats, and pigs that swarmed over the island in the wake of human settlement. And so it goes with island extinctions: the same story can be told countless times. All that differs are the names of the deceased and, sometimes, the identity of the perpetrator.

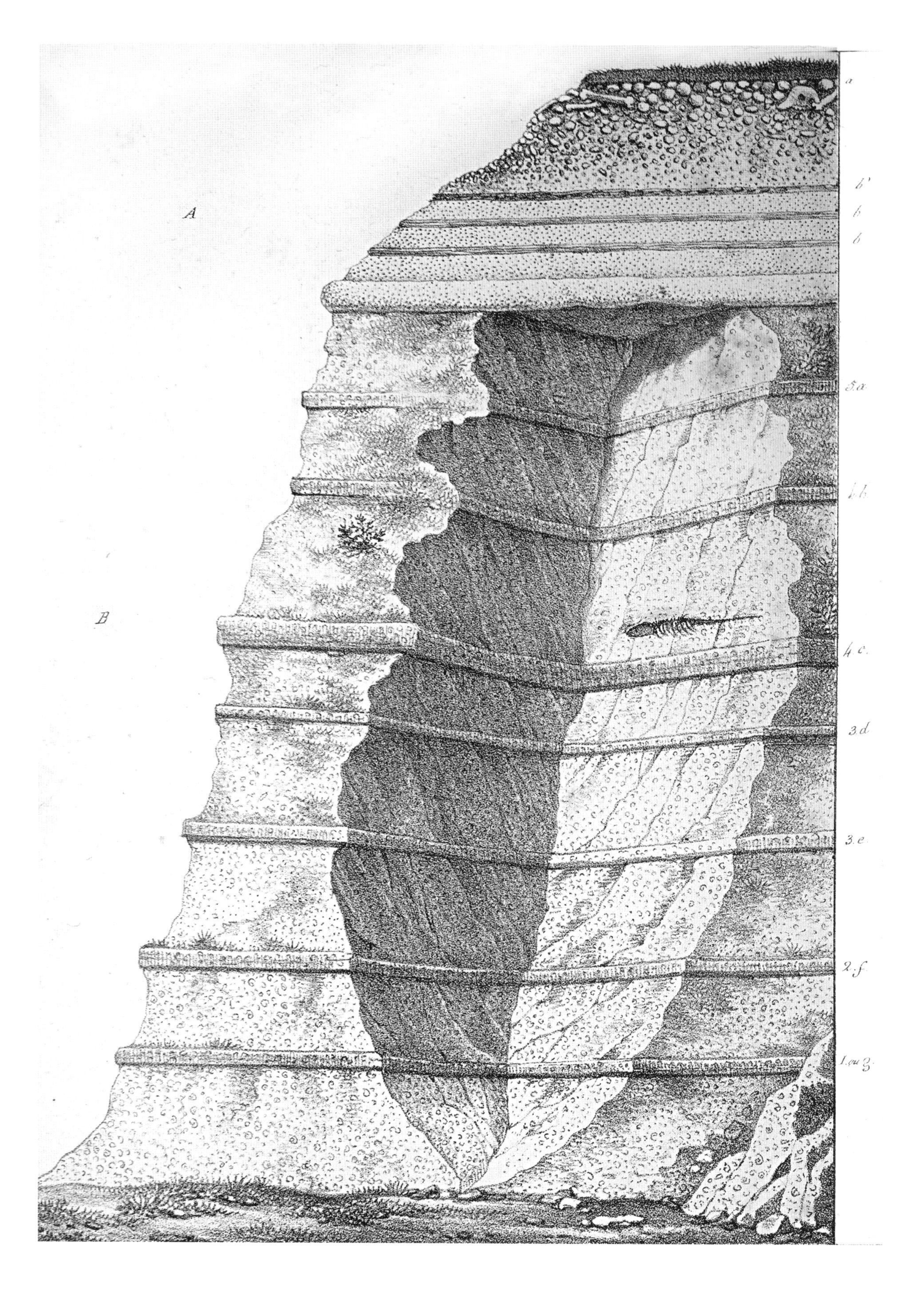

a
b'
b
b
A
5.a
4.b
B
4.c.
3.d.
3.e.
2.f.
1.ou 3.

5

EXPLAINING NEAR TIME EXTINCTIONS: FIRST ATTEMPTS

FIG. 5.1. **CUVIER'S GEOLOGICAL REVOLUTIONS:** During the course of his long career, Georges Cuvier conducted detailed geological surveys of the Paris basin, a large crustal depression in northern France filled with sediments of many different ages. By comparing how successive strata differed, often radically, in their composition and fossil content, Cuvier became convinced that the Earth must have been periodically affected by cataclysmic environmental change, or "revolutions." These revolutions resulted in great turnovers in species, as old ones died out and new ones came in from elsewhere. Note the skeleton of a large marine vertebrate exposed midway down the cliffside in this idealized section of a portion of the basin, as well as a skull and long bones, apparently of a proboscidean, perched just beneath the cliff edge. In Cuvier's view, such associations proved that sea and land must have alternated in this area, that revolutions had indeed occurred.

THE MAJOR THEORIES: CLIMATE CHANGE AND OVERKILL

Ideas usually have histories; good ideas usually have very long ones. Although fossils of extinct organisms were certainly encountered, and sometimes even collected, by ancient peoples, their significance was not understood until much later (see box, page 69). Even within fairly recent times, when biological extinction in the modern sense was considered at all, interpretations concerning causation usually involved vague appeals to "natural" change. However, by the mid-eighteenth century, some thinkers were already wondering about human persecution as an alternative or contributory cause of loss. In the 250 years since then, the relative level of support for each kind of explanation has swung fairly dramatically. That is still very much the case. In recent years, the human complicity argument has been ascendant, and how it came to be that way is best understood through the lens of history. (Those interested in more detail than I provide here can find a number of useful sources in the notes at the end of this book.)

THE RISE AND FALL OF ENVIRONMENTAL CATASTROPHISM

Prior to the eighteenth century, scholars interested in natural history (or natural philosophy, as it was then called) had to work under twin burdens. One was how to make intellectual sense of the emerging paleontological record, which included many forms of life that had no apparent analogs in the modern world. The other was to make that record, in all its complexity, jibe with accepted teachings from classical sources and religious texts. As Enlightenment ideals took progressively stronger hold on intellectual inquiry in the latter part of the eighteenth century, supernatural or esoteric explanations began to be discarded in favor of ones derived through observation and, increasingly, testable theory. But it was a long slog.

A good place to begin is with Georges Cuvier's concept of catastrophism, one of the most influential ideas in the development of early modern natural science. Cuvier (1769–1832) famously held that Earth's history had been marked by lengthy intervals of quiescence punctuated from time to time by stupendous cataclysms ("revolutions") that resulted, among other things, in enormous biological losses (see figure 5.1).

> Life, therefore, has been often disturbed on this earth by terrible events—calamities which, at their commencement, have perhaps moved and overturned to a great depth the entire outer crust of the globe. . . . Numberless living beings have been the victims of these

CYCLOPS AND THE DWARF ELEPHANT—A PALEOFANTASY

Even if they didn't know quite what to make of them, ancient peoples must have come across megafaunal fossils from time to time, perhaps in the form of strange-looking rocks or peculiar shapes in the ground. The myth of the Cyclops might just be based on someone's effort to explain the inexplicable, long ago. In classical mythology, the Cyclopes were a race of giants whose outstanding feature was their single eye, centered just above their noses. This seems a bit weird, even for Greek mythology. Was there something in nature that could have been the basis for the story? How about Mediterranean dwarf elephants? If you squint—a lot—you might be able to see a vague resemblance between the skulls of humans and these tiny proboscideans. Admittedly, the resemblance is far from detailed, and tusks, although reduced in the dwarves, are a bit of problem. But people see what they see: the large central hole must have accommodated a single, enormous eye—what else could it be for? In fact, the hole is the nasal aperture, which in all proboscideans is displaced high on the head in order to accommodate the base of the trunk. Unlike human eye sockets, which are encircled by bone and thus very obvious, in proboscideans they are unenclosed and therefore easily missed. Did the chance discovery of a dwarf elephant skull, perhaps by a Greek shepherd wandering into a cave looking for lost sheep, get transformed into a tale about a fearsome, one-eyed race of giants? And think how confusing it would be for a Cyclops, who had never seen people before, to come across a human skull!

Polyphemus the Cyclops, comparing a relative's skull to that of a strange-looking, two-eyed dwarf species he had just found. . . .

catastrophes; some have been destroyed by sudden inundations, others have been laid dry in consequence of the bottom of the seas being instantaneously elevated. Their races even have become extinct, and have left no memorial of them except some small fragments which the naturalist can scarcely recognise.

Cuvier noted that good fossil records show that many groups once occurred in areas far removed from the ones they or their relatives occupy today, in what were clearly quite different environmental circumstances. Since these groups are no longer where they once were, or even anywhere at all, something must have happened, something he judged to have been truly catastrophic (see figure 5.2). Fossil organisms found in rock records before, but not after, such revolutions must have either disappeared entirely or left the scene for other quarters. By the reverse token, species that made a sudden appearance within a record had to have come from somewhere else, greatly augmenting their ranges during the same catastrophes that caused the disappearance of other species.

Cuvier did not attempt to resolve the logical paradoxes embedded in some of his ideas. Although he was one of the first scientific thinkers to embrace the idea that complete extinctions do occur, he also accepted (or at least did not deny) that the number of species on Earth

ENVIRONMENTAL CATASTROPHISM

FIG. 5.2

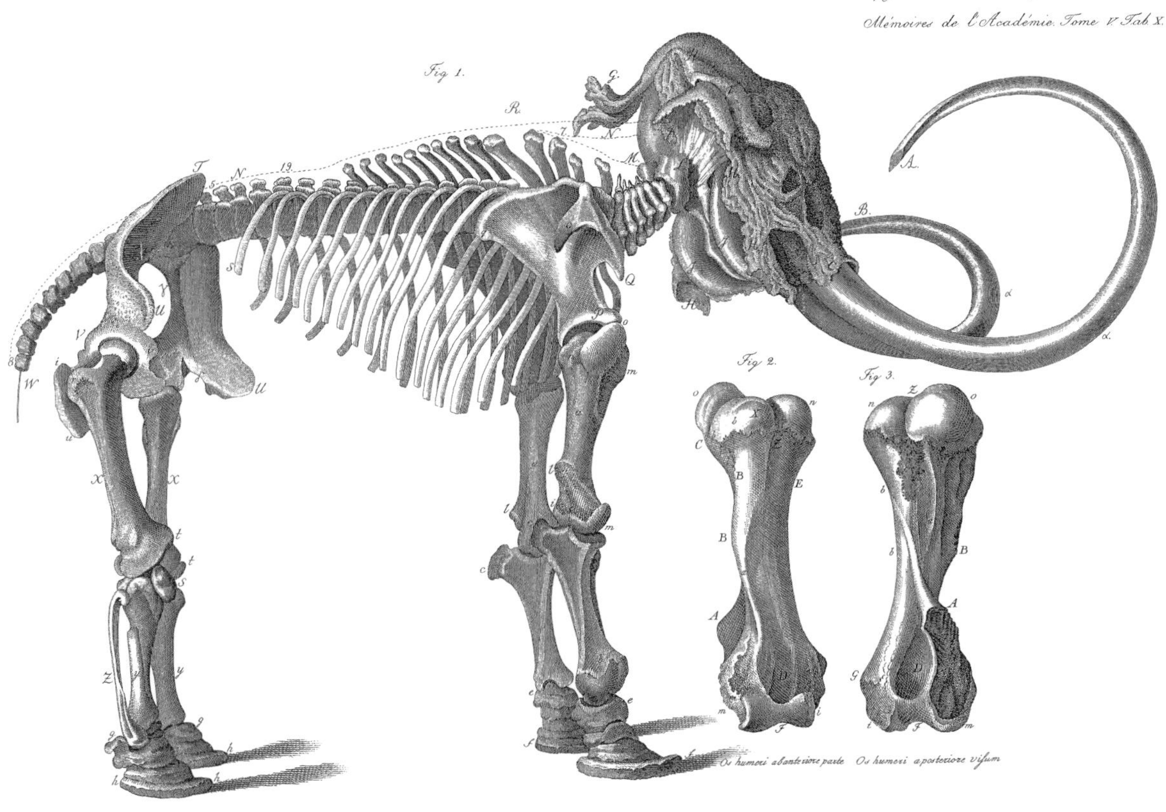

FIG. 5.3. **THE ADAMS MAMMOTH,** found near the Lena delta on the Arctic coast of Siberia, is one of the most complete specimens of an adult woolly mammoth that has ever been recovered. The mounted skeleton, with soft tissues still adhering to parts of the skull and feet, may be seen today in the Zoological Museum of the Russian Academy of Sciences in St. Petersburg.

had been fixed at the time of the biblical creation. But if extinctions had occurred, then the Earth's biota must have undergone widespread reductions during the revolutions he had detected in the geological record. Having no theory of evolution that could explain how new species could be constantly produced over time, Cuvier could only sidestep such contradictions by noting that, since so little was known about the Earth's former denizens, its past was still very much a mystery.

However, the French natural philosopher did provide an explicit mechanism for the disappearance of one species of interest here, the woolly mammoth. He had seen reports concerning the discovery of large numbers of mammoth fossils in Siberia, including well-preserved permafrost mummies such as the Adams Mammoth (see figure 5.3). These discoveries suggested to Cuvier that the species now known as *Mammuthus primigenius* must have been a casualty of one of his great calamities—perhaps dying out in a sudden deep freeze, which helpfully had the effect of preserving their carcasses. He wrote, "The same instant that these animals were bereft of life, the country in which they inhabited became frozen. This event was sudden, momentary, without gradation."

Although the idea of flash-frozen mammoths is no longer tenable, in tying their extinction to a unique kind of environmental change, Cuvier displayed sharp insight. Religiously inclined scholars, greatly impressed by Cuvier's views on Earth history, tried to turn catastrophism into a validation of scriptural accounts of calamities such as the Great Flood. Increasingly, however, such uncritical efforts were at odds with the torrents of new information coming in from all quarters of the planet (see box, page 73). Evidence of new lands, new peoples, and new species, lying entirely outside of any biblical frame of reference, established once and for all that looking to ancient authorities for comprehensive interpretations of the history of life was pointless. Henceforth, no endeavor would qualify as scientific unless it involved the gathering together of empirical facts in a form that could be critically evaluated and converted into explanations testable by still other facts—the same process that actuates disciplines in historical biology like paleontology today.

An example of this is Louis Agassiz's investigation of the biological effects of climate change caused by glaciation in the Northern Hemisphere, which was still catastrophist in outlook but much more firmly founded in geology. Agassiz (1807–1873), a Swiss-born naturalist, reasoned that the discovery in northern Europe of species now known only from tropical regions, such as extinct hyenas and rhinos, meant that higher latitudes must have once been much warmer than they were in his day (see figure 5.4). But then, as Agassiz later described it, a "sudden intense winter, that was also to last for ages, fell upon our globe . . . so suddenly did it come upon them that they were embalmed beneath masses of snow and ice, without time even for the decay which death allows." The winter that lasted "for ages" was of course a true ice age, one that persisted for tens of millennia, as Agassiz was among the first to establish.

Agassiz's recognition of the effects of the Ice Age on both biotas and landscapes was a triumph of observation and inductive reasoning. It pointed to a past catastrophe of correct Cuvierian proportions, but with less mayhem; for that reason, it was greatly influential in biological as well as geological thought. However, as the nineteenth century wore on, the influence of scientific catastrophism began to decline sharply. The cause was the development of a fundamental reinterpretation of the time scale upon which natural processes actually worked. James Hutton (1726–1797) and later Charles Lyell (1797–1875) proposed that the rock record was not produced by geological revolutions interspersed within long periods of little or no change, as Cuvier and the other catastrophists believed, but instead by the same slow, inexorable processes we see all around us today—the work of factors like water and wind, not overnight megafreezes or megafloods. Dubbed uniformitarianism, Lyell's

HUGEST, FIERCEST, STRANGEST! COME SEE, COME SEE!

When fossils of the American mastodon were first discovered, natural philosophers didn't know what to make of them, so quite understandably they dubbed their unknown owner "the American *incognitum*" (i.e., unknown thing). Charles Willson Peale (1741–1827), the famous American artist and naturalist, saw financial opportunity in the *incognitum*. Having acquired most of a mastodon skeleton unearthed in Ulster County, New York, he and his son Rembrandt did their best to put it back together, displaying the questionable result in their Philadelphia museum in 1801. The *incognitum*, it was asserted, must have been as "huge as the Precipice, cruel as the bloody panther, swift as the descending Eagle and terrible as the Angel of the Night." Rembrandt later decided the mount would do better box office as a carnivore than a herbivore, so he reversed the tusks in their sockets so that they curled down, like two extraordinarily overgrown canine teeth, arguing they might have been used for "striking down small animals, or in detaching shell-fish from the bottom of rivers, or even in ascending their banks." Useful adaptations indeed—sort of like having built-in ice axes that could double as an oyster pry. Although the Peales were clearly showmen, not biologists, the takeaway message is that accurately inferring the behavior of fossil animals can be difficult, especially when there are no close living relatives. That's why scientists interested in reconstructing behaviors of extinct species use an array of independent methods to make their inferences—not just wild imagination.

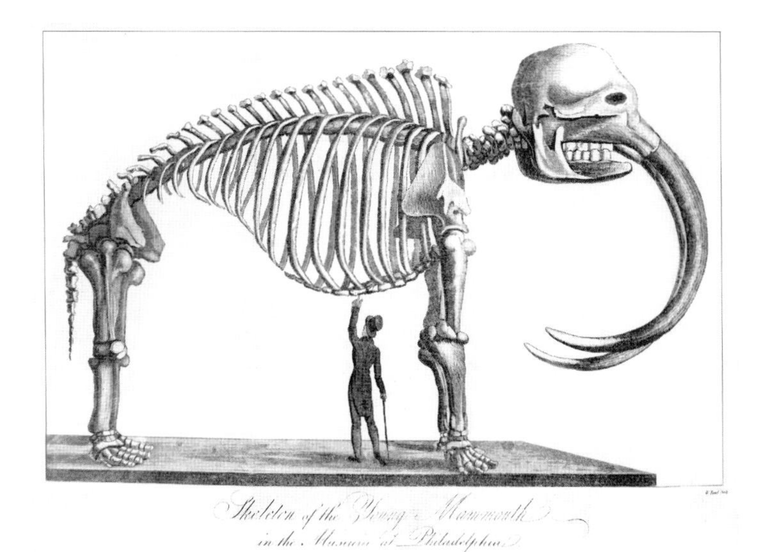

Peale's skeleton of the *incognitum*, arranged for maximum effect with minimum science. It was not until the early nineteenth century that mastodons and mammoths were clearly understood to be different kinds of proboscideans—hence the evident error in both name and spelling on the label ("mammouth").

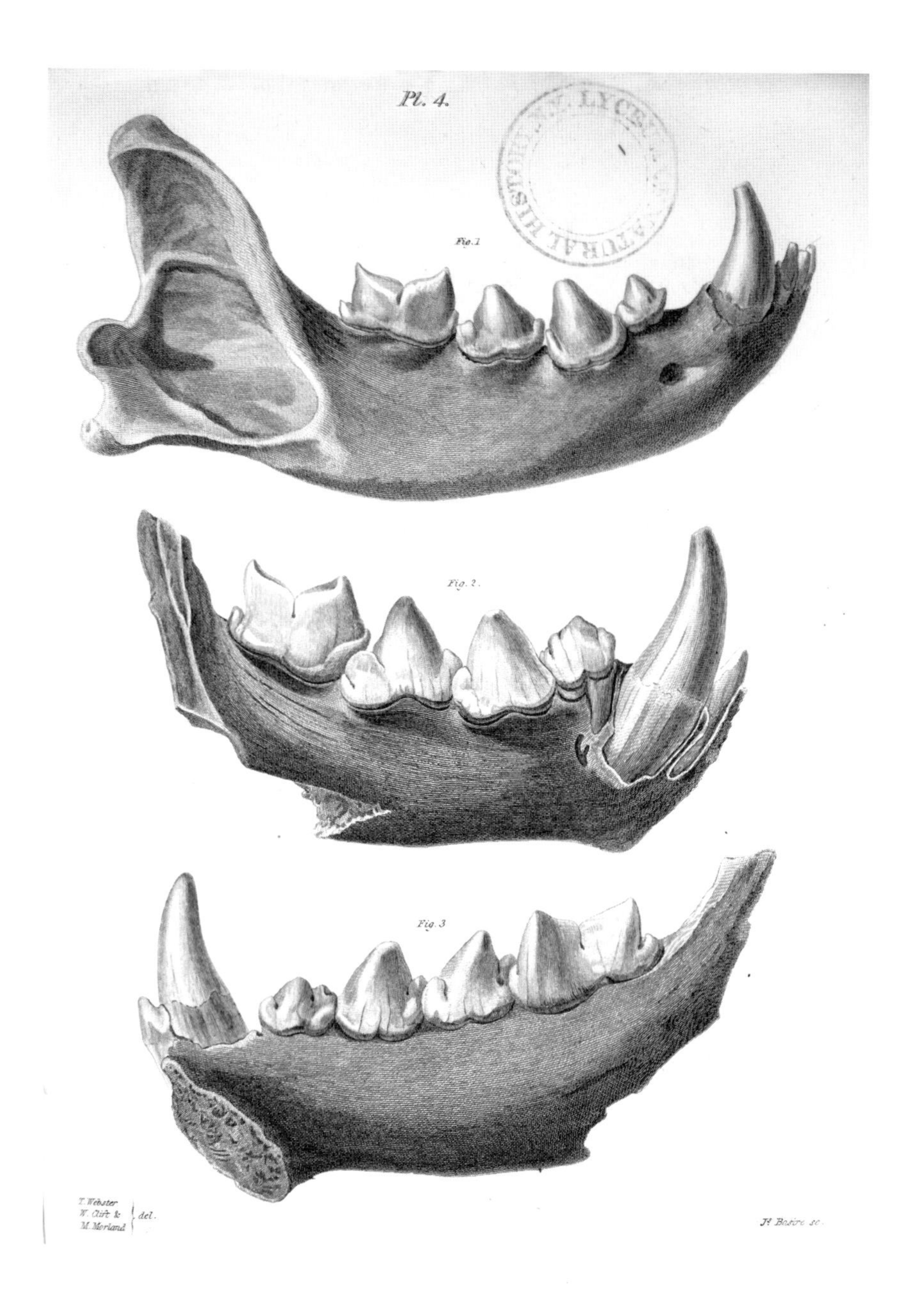

FIG. 5.4. **HYENAS ONCE LIVED IN BRITAIN:** This plate, published by theologian and early paleontologist William Buckland (1784–1856), compares a modern spotted hyena jaw (top) to that of a Pleistocene fossil from the site of Kirkland Cave in Yorkshire. They differ in size but are otherwise very similar, suggesting that Britain had once been much warmer than it is today. In addition to hyenas, Buckland also found remains of rhinos and hippos. A thoroughgoing but thoughtful supporter of Cuvier, he was convinced that there had been a worldwide inundation, but doubted that the floodwaters could have moved the bones of animals from the tropics to Britain. Later work showed that the spotted hyena disappeared from Britain and Europe at the end of the Pleistocene; it is now restricted to sub-Saharan Africa.

view that the present was the key to the past dominated later thought in geological and biological sciences, including the very new science of evolution.

In line with his geological views, Lyell thought that biological extinctions were mostly, if not entirely, due to the very gradual elimination of species, occasioned by equally gradual climate change. Charles Darwin (1809–1882), strongly influenced by Lyell's views, agreed that extinction was a gradual process, but emphasized the effect of contributory biological factors like natural selection and interspecific competition. He was more interested in laying a basis for understanding the effects of environment in driving evolution than he was in its possible role in extinction, but he certainly recognized that the two were inextricably related in their importance for the history of life. From Darwin's perspective, extinction could result in the complete annihilation of a lineage, or that lineage could evolve into something else, a new species. However, he did not endorse the concept of large simultaneous extinctions because the loss of many species all at once smacked of catastrophism, which he opposed in principle. Although he had to account for the wholesale disappearance of certain groups in deep time, such as trilobites and dinosaurs, he did this as a strict Lyellian, assuming that their extinctions were spread out over considerable time spans. Further, he argued that, although it may seem from the evidence of discontinuous geological sections that the species composing these groups disappeared simultaneously, they probably hung on here and there in areas or intervals as yet inadequately explored before finally succumbing. Possibly there was "some great system acting over the whole world" that influenced who lived and who died.

HUMANS AS A CATASTROPHE

As already noted, the belief that humans were somehow involved in Near Time extinctions is nearly as old as the idea that they were due to climatic factors. However, nothing amounting to a real inquiry into human complicity appeared until much later. In part the delay was due to religious sensibilities, which were not in accord with the notion that humans and long extinct megafauna could have been contemporaneous (see figure 5.5). Once human remains were found in unassailable association with extinct animals, however, attitudes began to change.

Ironically, it was the archgradualist Lyell who provided some of the first persuasive arguments in favor of human involvement in Ice Age extinctions. Although he had formerly doubted the coexistence of ancient humans and megafauna, by the 1860s the evidence that they had overlapped in time was incontrovertible. In his *Geological Evidences for the Antiquity of Man* he accepted this, claiming that:

> We may presume that the time demanded for the gradual dying out or extirpation of a large number of wild beasts which figure in the Post-pliocene [i.e., Pleistocene] strata, and are missing in the Recent fauna, was of protracted duration, for we know how tedious a task it

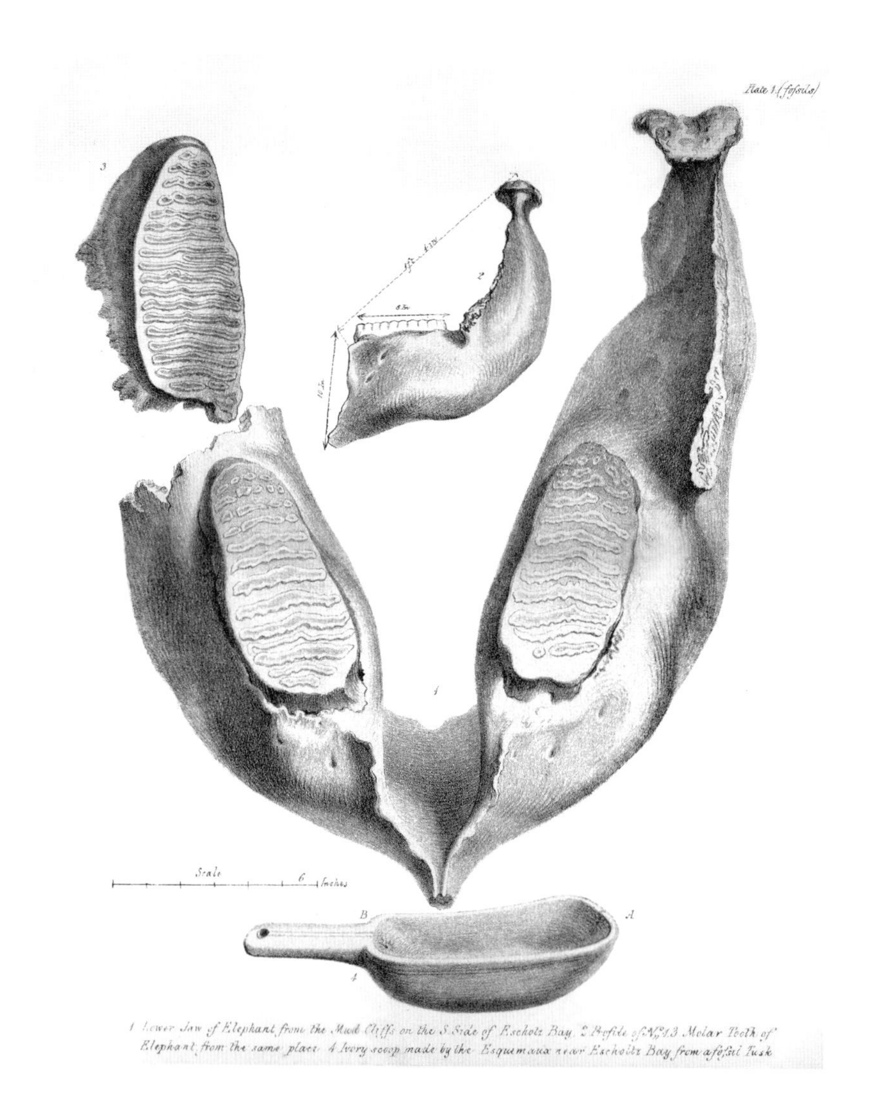

FIG. 5.5. **AND ELEPHANTS AND SLOTHS HAD EXISTED IN THE ARCTIC:** The first megafaunal remains from Arctic North America to be properly described were collected by a British expedition that briefly landed at Escholtz Bay on the west coast of Alaska in 1826. The bones were given to William Buckland, of cave hyena fame, to identify. He correctly recognized the bones and teeth of all the usual suspects—mammoth, bison, musk ox, horse, and reindeer—but even he was stumped by one specimen. It was a vertebra, probably that of Jefferson's ground sloth, *Megalonyx jeffersonii*, now known to have lived in the far north during interglacial times. Buckland thought that the extinct mammals must have died out in a sudden deep freeze: it would be "vain to contend that they have been subdued and extirpated by man, since whatever may be conceded as possible with respect to Europe, it is in the highest degree improbable that he could have exercised such influence over the whole vast wilderness of Northern Asia, and almost impossible that he could have done so in the boundless forests of North America." The artifact illustrated in the plate, an "Ivory scoop made by the Esquimaux . . . from a fossil tusk," was acquired in trade and was of recent manufacture.

is in our own times, even with the aid of fire-arms, to exterminate a noxious quadruped, a wolf, for example.... That the growing power of Man may have lent its aid as the destroying cause of many Post-pliocene species must, however, be granted.

Still, his was not a pure overkill argument. In Lyell's view, persecution by humans was a contributory rather than an exclusive cause of megafaunal extinction. Nor would it have happened quickly, he contended:

It is probable that causes more general and powerful than the agency of Man, alterations in climate, variations in the range of many species of animals, vertebrates and invertebrates, and of plants, geographical changes in the height, depth, and extent of land and sea, some or all of these combined, have given rise, in a vast series of years, to the annihilation, not only of many large mammalia, but to [that of others as well].

Having added the rhetorical equivalent of the kitchen sink to the list of probable causes of megafaunal extinctions, Lyell must have considered the question of what *precisely* caused biological losses in any given instance to be either beyond human comprehension or too obscure to be answerable.

Perhaps surprisingly, Darwin took no strong position regarding whether or not human activities had been a major cause of recent extinctions. As University of Washington archeologist Don Grayson has noted, Darwin was of course aware of Lyell's eventual espousal of human depredations as an explanation for megafaunal losses in Europe and North America. However, in the end, Darwin just didn't know why some extinctions occurred and said so, writing, "We need not marvel at extinction: if we must marvel, let it be at our presumption in imagining for a moment that we must understand the many complex contingencies, on which the existence of each species depends."

The contribution of Alfred Russel Wallace (1823–1913) to evolutionary biology is sometimes overlooked because he worked in the shadow of his much more famous contemporary, Charles Darwin. Wallace was a man of many parts, some of which must have been in violent disagreement with one another. A brilliant observer, he also believed in angels. Near the end of his life he developed the idea that evolution was in fact directed, not by natural selection as he had once thought, but by those very same heavenly messengers. But he was also a man who wrestled with big questions, and he was considerably more insightful than many of his contemporaries regarding megafaunal extinctions:

We cannot but believe that there must have been some physical cause for this great change; and it must have been a cause capable of acting almost simultaneously over large portions of the earth's surface.... Such a cause exists in the great and recent physical change known as "the Glacial epoch"... [which] must have acted in various ways to have produced alterations of level of the ocean as well as vast local floods, which would have combined with the excessive cold to destroy animal life.

Although this passage gives the impression that Wallace's thinking about causation was really not much advanced over that of Cuvier and Agassiz, in fact he thought that the onset of "the Glacial epoch" explained only certain losses. He could imagine how glaciation might have had an effect on animals living at higher latitudes in the Northern Hemisphere and also southern South America, which likewise bore traces of Pleistocene glaciation and had by this time yielded a number of extinct megafaunal species. What he couldn't understand was how high-latitude glaciation could have severely influenced the more tropical parts of the world such as Australia, where recently extinct species "rivalling in bulk those of the great continents" had been described by the influential anatomist and paleontologist Richard Owen (1804–1892), who had begun to wonder about a human role in these extinctions (see plate 4.10). Wallace had no answer, but he was asking the right kind of question. Something else had to have been going on.

Quaternary paleontology matured into a major discipline in the early twentieth century. Time had become an important parameter for all paleontological inquires, although there was still no way to measure it very accurately. But by the early 1930s there were some inklings about the confluence of people, time, and megafaunal extinction, as Alfred S. Romer (1894–1973), a prominent Harvard paleontologist, noted:

> [The] overwhelming trend of the evidence is that very little extinction seems to have taken place [in North America] during the Pleistocene proper, and that a vast amount of extinction, reducing the fauna to its present impoverished condition, has taken place in a comparatively short period which presumably cannot have had its initiation more than, roughly, 20,000 years or so ago.... That [*Homo sapiens*] played any major direct part in actually killing off this fauna is far from probable, for in that case, we should find much more evidence of his association with extinct types than exists at present. But it might be suggested (very tentatively) that the appearance on the scene of a new form of this sort might possibly have thrown out of balance a fauna in delicate adjustment, and indirectly have caused considerable changes.

Henry Fairfield Osborn (1857–1935), an early president of the American Museum of Natural History and a leading mammalian paleontologist, observed that "in Eurasia, man 'grew up' with the Pleistocene faunas of that region, so that he was always more or less in a state of ecological balance with the mammals around him." By contrast, humans were new in, and late to, North America:

> [I]t may have been the entrance of this destructive animal, even though not at first important, that caused the extinction of so many great mammals.... Not that man was directly

> responsible for the killing off of numerous and large herds of mammoths, horses, and camels, but rather by his entrance he may have upset an ecological balance, he may have introduced epidemics, or he may have contributed in other ways which from this distance are quite obscure.

And everything might very well have remained just as tentative and obscure as Romer and Osborn said, had it not been for the invention of radiocarbon dating in 1946, which in fact did change everything. With substantial improvements in instrumentation and effective range (roughly equivalent to the 50,000 years of Near Time), by the 1960s radiocarbon dating could be used to determine the age of anything organic that retained a measurable amount of radioactive carbon, with a degree of precision and accuracy unachievable by any other method.

And radiocarbon also changed everything for a young paleoecologist working at the University of Arizona, Paul Martin (1928–2010). Already deeply involved scientifically with the world of the Pleistocene by the late 1950s, he realized that radiocarbon dating was the way to demonstrate unequivocally whether potential causes of extinction actually preceded their presumed effects, and how closely. With regard to Near Time extinctions, it was now feasible to test the temporal correlation of losses that had occurred in different parts of the world. And, of course, radiocarbon could also date recent chapters of the human diaspora. It was time to propose a new common-cause explanation, one that had planetwide application and a serious testing procedure.

6

PAUL MARTIN AND THE PLANET OF DOOM: OVERKILL ASCENDANT

FIG. 6.1. **A MAMMOTH HUNT,** as imagined by Charles Frederick Holder, a curator at the American Museum of Natural History in the 1880s. Long before the appearance of the overkill hypothesis, Holder had already envisaged that the "one agency that might have produced [the mammoth's] extermination is . . . man. There is little doubt that the early Americans chased the great animal, and, hunted from one part of the country to another, they finally disappeared." Although there is nothing particularly realistic about this scene, it makes the self-evident point that elephant hunting could be dangerous. Picturing mammoths as oversized African elephants with retroverted tusks was a common (and inaccurate) artistic convention at the time (compare plate 7.1).

MARTIN LOOKS FOR THE KEY

Although by the turn of the twentieth century it was increasingly suspected that humans were implicated in Quaternary disappearances, no one had been able to nail down what might have happened in any precise way. Alfred Romer could only guess that some species existed "within one or two score thousands of years of modern times," but that "a number of forms may have lingered on to within a few thousand years, or even less, of modern time." Given so porous a time line, when anything could have happened at any time, there was no saying what had been responsible for the megafaunal losses, or indeed whether there had even been a single cause. Paul Martin's contribution was to finally give some rigor to the extinctions dating game by emphasizing the implications of the radiocarbon record. He first showed that there seemed to be a startlingly close temporal relationship between human arrival and animal extinction in North America. Combing the literature, he found that the limited radiocarbon databases then available for other areas of the globe also appeared to support, or at least did not contradict, his basic contention, that when the humans come, the animals go. On this basis, he argued that the apparently tight correlation between human arrival and species disappearances had been a worldwide, repeating phenomenon, one that seemingly took place within decades to centuries in the case of many islands, or a millennium or so at most in the case of continental areas. Even widespread megafaunal species like mammoths and ground sloths seemed to have declined rapidly, dying out wherever they lived within a very brief period (see box, page 83). Could even the most extreme forms of climate change have caused such accelerated losses? he asked.

Martin had a correlation, but he needed a cause-effect relationship. If humans were to be decisively implicated in these losses, an acceptable kill mechanism still had to be identified. For example, some early authors speculated that early human migrants to Madagascar had caused disastrous landscape alteration through unrestricted burning, resulting in the loss of species dependent on forested habitats. However, there was no evidence for this kind of environmental devastation in places like high-latitude North America and Eurasia, where substantial losses occurred nonetheless. Just being on the scene was not enough; prehistoric peoples had to have done something that culminated in massive extinction, something within their technological and behavioral capacities. For Martin there was only one universal human practice that fit these requirements: predation, or, as we prefer to style it when we are speaking about ourselves, hunting (see figure 6.1).

For almost all of our history as a separate species, our ancestors lived by hunting and gathering—that is, by acquiring their food and other necessities directly from nature, without any form of stable agriculture or animal husbandry. Therefore, according to Martin, if prehistoric humans had caused the loss of hundreds of species in Near Time, this would probably have involved hunting in some form. But could ordinary hunting, as practiced by modern hunters and gatherers, do the job? There was no indication in ethnographic sources

WRANGEL ISLAND, THE WOOLLY MAMMOTH'S LAST STAND

Wrangel Island is a surpassingly isolated place. Ten thousand years ago it was part of the Siberian mainland, just a promontory jutting into the Arctic Ocean at the edge of the exposed continental shelf. As sea level began to recover, the promontory became an island. For one small population of mammoths, this was providential: separated from events unfolding on the mainland, they managed to survive there until 3,700 years ago.

When radiocarbon expert Sergey Vartanyan and his colleagues began their work on the island in the late 1980s, all that was known about Wrangel mammoths was that their bones and tusks littered the island's tundra in huge numbers. As a result, the scientists were able to collect and date numerous samples—with surprising consequences for extinction studies. Initially, Vartanyan's radiocarbon results seemed unbelievable: they were thousands of years younger than the next oldest, and thus were in conflict with the accepted view that mammoths had disappeared from Eurasia by the beginning of the Holocene. His dates told a different story, of unexpected survival at the very edge of the Pleistocene world. Soon other researchers were recovering similar radiocarbon ages on Wrangel fossils, ruling out the possibility that Vartanyan's original dates had been a fluke or the result of contamination.

Mammoth tusk sample, collected on Wrangel Island for radiocarbon dating.

PLATE 6.1. **AMERICAN FALSE CHEETAH RUNNING DOWN A PRONGHORN:** The cougar (*Puma concolor*) and the jaguar (*Panthera onca*) are the only large cats remaining in North America, but at the end of the Pleistocene there were several others, including the American false cheetah, *Miracinoyx trumani* (with an estimated body size of 70–95 kg/140–190 lb). Its common name reflects the fact that, according to ancient DNA investigations, it is actually more closely related to the cougar than the cheetah despite what its original describer though. This cat's range stretched from Pennsylvania to Florida in the east and to the Grand Canyon in the west. Pumas and jaguars are ambush predators, and in all likelihood so was the false cheetah. Skeletal remains indicate that it possessed long legs and wide nostrils to facilitate rapid inhalation and exhalation, which are features expected in a speedy pursuit carnivore. In the illustration *Miracinoyx* is pursuing an extinct pronghorn, *Tetrameryx*; however, its overall diet was probably much like that of living cougars, which hunt everything from mice to mammals twice their size.

that people following this way of life in recent times had ever caused the elimination of anything. The likeliest social unit for ancient, preagricultural peoples—the band—seemed too small, too loosely organized, to function as a killing machine on a large scale (see chapter 9). Besides, ethnography showed that most hunter-gatherers were much more reliant on foods that didn't move around much—fruits, tubers, berries, shellfish, occasionally insects—than they were on the few (usually small) birds or mammals that they might actively hunt. Martin therefore had a problem, which he attempted to solve by reimagining the nature of human hunting within the unique circumstances attending the arrival of people in new lands for the very first time.

In typical mammalian predator-prey relationships, the hunter and the hunted are bound together by instinct as well as learned behavior as performers in the oldest interactive game on Earth. In a way, this relationship is as intimate as a love affair. The lynx not only recognizes the snowshoe hare as its preferred target, it knows in its marrow the kinds of feints and maneuvers the hare will make to avoid being caught once the chase is on. In the same sense, the lion knows the zebra, the wolf the caribou, the owl the mouse, and so on (see plate 6.1). But rarely, if ever, does this result in unrecoverable levels of loss on the part of the prey because of overhunting by the predator. This only makes sense. As the game is played out in real life, attacks are often unsuccessful; each side knows the other's likeliest moves, but cannot control for all exigencies. Nor would it be in the predator's long-term interest to increase its hunting efficiency, even if it could, to the point that its preferred prey ends up being threatened by extinction. The predator might live in plenty for a while, but if its usual target disappears, it must either find other, perhaps less suitable, targets or starve. Thus in a normal predator-prey relationship selective forces operate in a sort of dynamic equilibrium, with both sides benefiting from having the less fit weeded out. The scales might tip toward disaster from time to time, as in the case of island contexts where predator and prey fail to achieve equilibrium. But such anomalies are likely to be localized and transient, with no long-term consequences for species that enjoy extensive ranges, as do most large herbivores and the predators that depend upon them. The challenge to Martin's hypothesis was therefore obvious: how could human predation possibly account for the pattern of extinctions in North and South America, which involved dozens of continent-wide species losses within absurdly short periods of time?

Martin's response was to insist that there was one circumstance in which typical predator-prey interactions might be disabled for a time, so that neither party would act in the manner expected. This may occur when a new, very capable predator enters an environment in which it had not previously been present. Potential prey, suddenly vulnerable, would lack inherent fight-or-flight behaviors attuned to the new arrival. The interloper, cloaked by its singular differences from anything in the prey's experience, would trip no antipredator defenses, raise no alarms. In such circumstances, Martin argued, prey species would have acted naïvely, not recognizing the existence of a novel threat until it was far too late.

Meanwhile, the predator could act without limit, taking down anything and everything encountered, until at some point the prey's learning through experience or its increasing rarity ended the period of hypersusceptibility. In the extreme scenario that Martin had in mind, in which the newly arrived predator is no ordinary exotic but instead culture-bearing, unimaginably dangerous *Homo sapiens*, most species would not have had sufficient time to react or to learn (see box, page 87). The handful of large-bodied species lucky enough to escape or survive biological first contact, as it is called, were the exceptions that proved the rule.

In Martin's mind, he had now accounted for everything important: no coordinated losses until humans come along, then no respite until overhunting had run its inevitable course.

BIOLOGICAL FIRST CONTACT AND PREY NAÏVETÉ

As his many critics noted, there was much that seemed either flatly wrong or highly improbable in Martin's overhunting hypothesis. Indeed, as a term, "overhunting" was far too mild a descriptor for what Martin himself had in mind, so he soon rebranded it as "blitzkrieg"—an intentionally haunting metaphor that suggests a relentless, almost fanatical devotion to annihilation, the application of overwhelming force against feckless, supine victims (see figure 6.2).

Yet reality differs from metaphor. To be sure, there are numerous modern examples of human persecution in which individual species have experienced serious harm in the form of range collapse, disruption of breeding cycles, and consequent major population losses. But under what credible circumstances could peoples with unsophisticated tool kits maintain exceptional levels of hunting pressure, on not one but many species simultaneously, in a manner sufficient to cause the utter collapse of most, or sometimes all, of them?

Martin's scenario does not lack basic credibility. It bears some resemblance to the phenomenon of ecological release, well attested in many kinds of settings, whereby an invasive species literally bursts onto the scene and manages to supplant or otherwise seriously imperil members of the local flora and fauna. Ecological release can theoretically result in the extinction of resident native species, and has probably done so in any number of interactions in island contexts, but the usual outcome is that the invader eventually settles into a suitable niche and is thereafter not much of a bother to others. To cite an invertebrate example, the European zebra mussel (*Dreissena polymorpha*) has been spreading through waterways in the eastern United States and Canada since its introduction in the late 1980s, causing considerable economic as well as ecological damage because it tends to coat every available surface in lakes and rivers with abundant examples of itself. Although certainly a threat to endemic

CAVE ART: THE ANIMALS ON THE WALLS

The most arresting evidence of long-term human–megafaunal interaction comes not from kill sites and the remains of ancient meals, but from art, in the form of images painted (as pictographs) or carved (as petroglyphs) onto cave walls. These were made by people whose mental processes must have resembled ours, but whose life experiences and outlook would surely have been very different. The best-known examples of cave art are in France and Spain, but pictographs of various ages have been discovered in many other locales as well. European pictographs are thought to mostly date to a period 30,000 to 40,000 years ago, some 20,000 years or more before the end-Pleistocene extinctions. Cave paintings often reveal in gorgeous detail the external form, coat colors, and even the expressions of the animals whose natures the artists were trying to capture. One theory is that their makers believed that such depictions would have possessed magical properties, perhaps helping to ensure successful hunts. In other cases, the artists may have just been doodling—or drawing graffiti, if you prefer. After all, they were human just like us, and scratching out an image or two on a cave wall during a long winter's night might have helped alleviate a dose of Paleolithic boredom.

Cave artists at work in Font-de-Gaume Grotto, Les Eyzies-de-Tayac, France, as imagined in 1920 by Charles R. Knight, whose paintings frequently depicted prehistoric life.

FIG. 6.2

freshwater mollusks and fishes because its huge numbers disrupt their food webs, and despite early fears, the zebra mussel does not appear to have induced any complete exterminations—at least not yet. A pest of the first order, yes, but probably not a kill mechanism. Faunal blitzkrieg, by contrast, is a perversely amplified version of such an alien invasion, causing truly serious harm in the form of multiple, concurrent extinctions of native species.

Martin argued that overhunting would have provided food security on a heretofore unachievable scale, enabling the early residents of the Americas to successfully raise many more children than would have been otherwise possible. With larger populations came increasing needs, with the result that overhunting became a self-propelling instrument for the maximum extraction of energy from the environment with an apparent minimum of effort. The most desirable prey species would have been the largest, since they would have provided the most substantial nutritional packages, fats as well as meat. This positive feedback loop presumably operated until the boom finally went bust, as megafauna disappeared completely or were no longer easily found. In the continental New World humans eventually settled down, like so many zebra mussels, into less damaging lifeways such as ordinary hunting and gathering and, in some areas, agriculture.

MULTIPLE SPECIES, MULTIPLE EXTINCTIONS

If Martin's view that the motivation to overhunt was an expression of an inherent drive to maximally reproduce seems unconvincing, this is hardly the only explanatory burden that his overkill hypothesis has had to shift. For example, if the ecological disasters associated with overhunting went to completion in a few hundred years, as the dating evidence seemed to indicate for the continental Americas, then this suggests that the hunters must have been essentially indiscriminate, going after many species more or less simultaneously. This is far out of the norm for technologically unsophisticated hunters, who tend to concentrate on a small number of species whose habits they know and can hunt at optimal times during the year. Martin was asking for a form of human behavior that had never been seen ethnographically. How could this unrelenting, steroidal form of hunting have worked?

In 1975 Martin answered this and related objections with a simulation model he developed with a collaborator, statistician James Mossiman. In later papers and his last book, *Twilight of the Mammoths*, published in 2005, Martin added details and discussed new evidence, but he never abandoned his fundamental sense of how faunal blitzkrieg had torn through the Americas. In its original form the model begins with early humans passing into the midcontinent roughly 12,000 calendar years ago along the newly open ice-free corridor (see chapter 4 and figure 3.3). Highly mobile paleohunters would have fanned out into new areas at lightning speed, surprising and felling their prey long before they had time to develop avoidance mechanisms. With devastation complete in one region, the humans would have moved on to the next in a perfect storm of destruction. The constant availability of great quantities of food would have encouraged rapid increases in population, which in turn would have prompted more migration, to areas as yet untouched. Simultaneous targeting of multiple species was thus transformed from a possible liability into a potential strength. It was simply what would have been expected under blitzkrieg conditions.

With no apparent limit to their mobility or natural rate of increase, in Martin's model early humans would have passed rapidly through North and Central America, attaining the tip of South America within a thousand years or so of original biological first contact, which presumably occurred in Beringia or the ice-free corridor. By the start of the Holocene, whatever remained of the megafauna in the Americas—moose, deer, elk, caribou, bison, mountain goats, bighorn sheep, musk ox, bears, guanacos, vicuñas, tapirs, peccaries—either had managed to avoid the paleohunters by living in remote places, had had enough time to absorb the danger posed by humans, or were just plain lucky. Like a grass fire burning itself out on the prairie, everything consumable had already been devoured along the way. With all susceptible species either already extinct or nearly so, the era of big-game overkill had come to an end.

7

ACTION AND REACTION

PLATE 7.1. **A MALE COLUMBIAN MAMMOTH** *(Mammuthus columbianus)* in the American Southwest: Mammoths differed from living elephants in various ways—it wasn't just a matter of all the hair. At least in adult males, the tusks were much larger relative to body size than in living African and Asian elephants. In mammoths, the tusks grew with a slight twist, so that they progressively crossed each other as seen here. Bodily proportions were also different: the huge head, small ears, and sloping back of mammoths are features not seen in living elephants. (Despite the connotations of the word "mammoth," the bodies of woolly mammoths were rather small compared to those of extant elephants.) Although the Columbian mammoth lived in the warmer, midcontinental part of North America, analyses of ancient DNA have shown it was able to hybridize with the woolly mammoth. How often this occurred is hard to say, but in living mammals natural hybridization among otherwise distinct species is known to occur (wolves, coyotes, and domestic dogs can all interbreed, for example).

CLIMATE CHANGE STRIKES BACK

Ironically, the overkill hypothesis has been at least partly responsible for reviving interest in climate change as a possible cause of Near Time extinctions. For many archeologists and Quaternary paleontologists, the assumptions of overkill—extraordinary levels of prey naïveté, implausible rates of slaughter, absence of kill sites—were and are unacceptable. As paleontologist John Guilday once put it, "To single out a particular predator or a set of circumstances is fun but futile."

A frequent objection is to the effect that, if climatic and other natural events had been overwhelmingly responsible for extinctions throughout the ages, why shouldn't that have been the case in the Late Pleistocene as well? We now know, in incredible detail, that we live in a world that has in the past experienced massive, short-term fluctuations in climate in ways that must have made life hard, if not impossible, for species living in unlucky places at fateful times. Adopting climate change as the driver does not obviate the need for special assumptions, as we shall see, but it does take aberrant human behavior out of the equation, and for many this was its fundamental appeal.

Although many authors have attempted to compose robust, climate-based alternatives to overkill, especially for the continental Americas and Australia, the results tend to have shared failings. Whereas Paul Martin offered an apparently global explanation for an allegedly global pattern of extinction, most climate change explanations hardly go further than the data sets used to develop them. Sophisticated scenarios have been developed to explain regional losses as the result of ecological transformations due to gradual climate change, but organisms with extensive ranges such as mammalian megaherbivores seem unlikely candidates for complete extinction under these conditions (see, for example, plate 7.1). At least in principle, they could have simply migrated to more favorable regions. Clearly, other factors had to have been involved.

A great deal of new paleoenvironmental evidence has become available in recent decades, but still, for many parts of the world, nothing has emerged in the way of identifiable natural causes that seem severe enough or general enough to have acted as extinction drivers at all necessary times and places. Just as with overkill, timing is everything. If atmospheric and oceanic circulation patterns in the Southern Hemisphere caused environmental change serious enough to force numerous losses in Australia 40,000 years ago, then why didn't latitudinally equivalent southern Africa suffer in the same way at the same time? (See plates 7.2 and 7.3.) If the Younger Dryas cold interval critically reduced megafaunal ranges in North America and Europe 11,700 to 12,900 years ago, thereby setting the stage for their final collapse, why didn't these species disappear during other, equally significant cold snaps earlier in the Pleistocene? In the Northern Hemisphere, environmental events that were unquestionably of enormous magnitude seem to have had no marked effect on the megafauna, except perhaps to reduce some population sizes and ranges in such a way as to

COEVOLUTIONARY DISEQUILIBRIUM

FIG. 7.1

force a few isolated disappearances. In fact nearly all of the well-dated megafauna in both Eurasia and North America weathered the Last Glacial Maximum, only to die out at a much later time.

For Russell Graham of Pennsylvania State University and Ernest Lundelius of the University of Texas at Austin, an overlooked factor was not gradual changes, which could come about for many reasons, without any necessary consequences, but instead the loss of ecological equilibrium at critical times. In an influential paper they proclaimed that "there are no modern analogues to Late Pleistocene biotas or environments" of North America, the evidence for this conclusion being the observation that the flora and fauna of that age

1. Giant short-faced kangaroos *(Procoptodon goliah)*
2. Giant diprotodont *(Diprotodon optatum)*
3. Wonambi snake *(Wonambi naracoortensis)*
4. Mihirung *(Genyornis newtoni)*
5. Kookaburra *(Dacelo novaeguineae)**
6. Pink cockatoo *(Cacatua leadbeateri)**
7. Ninja turtle *(Ninjemys oweni)*

[* = an extant species]

PLATE 7.2. **DRY WOODLAND PANORAMA, SOUTH AUSTRALIA:** This scene of rich diversity depicts, on the right, a pair of giant short-faced kangaroos sunning under a eucalyptus tree. There were several species of these kangaroos; the largest, estimated to have weighed around 240 kg (530 lb), was three times the size of the largest living kangaroo (red kangaroo, *Osphranter rufus*). Short-faced kangaroos seem to have been bipedalists rather than hoppers, trotting along on their single, hooflike nails. Also seen is a female giant diprotodont with her young. This species was the largest of all the marsupial megafauna and markedly sexually dimorphic (males were 2750 kg/6000 lb; females, half that size). The mihirung, a large, strongly built flightless bird, was a distant relative of ducks and geese. Its large, powerful beak may have been an adaptation to break apart hard foods. Creeping along in the rear is a large ninja turtle (plate 8.9), and in the dry stream bed lurks a wonambi (plate 4.4). An extant kookabura sits patiently on a limb above, waiting for a small lizard or other prey to make a wrong move. The pink cockatoo is extant and prefers extensive woodlands like the one depicted.

PLATE 7.3. **GIANT AFRICAN BUFFALO AND QUAGGA-COLORED ZEBRA IN SOUTHERN AFRICA:** *Pelorovis antiquus*, a relative of the living Cape buffalo (*Syncerus caffer*), has been tentatively identified at the archeological site of Jebel Irhoud in Morocco, where anatomically modern humans were hunting large mammals as much as 300,000 years ago. During the Late Pleistocene, the giant buffalo ranged through much of Africa, but by 12,000 years ago it had disappeared from the southern and eastern parts of the continent. Surprisingly, *Pelorovis* held out in northern Algeria until as recently as 4000 years ago. *Pelorovis* was a massive animal, possibly weighing as much as 1200–2000 kg (2400–4400 lb), which is the size of the living black rhino. Its most remarkable feature would have been its enormous horns, the tips of which may have spanned 2.5m (8 ft) according to some estimates. The other animal in the painting is a plains zebra, but it has an unusual coat color—striped in front as expected, but shading to a monochromatic brown toward the haunches. Until the mid-nineteenth century, this type of coloration (known as quagga) was common in some plains zebra populations. Then the coat color became increasingly rare; it has not been seen in the wild for the past 150 years. This is peculiar, because ancient DNA reveals that quagga-coated animals were not genetically distinct from other plains zebra. To add to the confusion, it was recently settled that the valid scientific name for the plains zebra is actually *Equus quagga*!

were "disharmonious" in their organization (see figure 7.1). In the context of their paper, this word had a technical meaning—namely, that species compositions were different from those seen today, which were judged to be "harmonious." This was a deliberate choice of terms, since a connotation of harmoniousness is that, whatever the grouping—biological species, highway traffic, instruments in an orchestra—the units under consideration are meant to work with one another in a concordant manner. Disharmony carries the opposite connotation.

Graham and Lundelius visualized communities of plants and animals as coevolved wholes. In long-standing biotas there will be a natural tendency toward ecological and evolutionary harmony, with plants and animals coexisting in balance. Selection will ensure that the herbivores do not overharvest and that the plants do not evolve antipredator defenses at a faster rate than the herbivores can tolerate. If boundary conditions remain relatively the same, coevolutionary relationships should as well. But climate change is the great disruptor, because it breaks up these stable relationships. Graham and Lundelius argued that Late Pleistocene environments were characteristically maintained by climates with low seasonality. That changed during the Pleistocene–Holocene transition, with the critical period lying between 11,000 and 13,000 years ago. Communities had to restructure because of lost equilibrium; the fallout from reintegration was the extinction of the more specialized megafauna. Species that had sufficient adaptability, because of their niche breadth, survived.

OVERKILL REDUX

The appeal to ecological theory, rather than to paleohunters gone wild, was welcome to many participants in the Near Time extinctions debate in the mid-1980s. At that time one could still argue that the interval between the end of the Last Glacial Maximum and the Pleistocene–Holocene transition was traumatic enough to force losses in places as far removed from each other as North America and Australia. However, scarcely any reliable radiocarbon dates were then available for the Australian extinctions, and the few that were tended to cluster at a point much earlier than the terminal dates for the North American megafauna. Since then, investigations utilizing OSL dating have convincingly demonstrated that most, if not all, Late Pleistocene extinctions in Australia occurred around 40,000 to 45,000 years ago (see plate 7.4). This was 30,000 years before the New World extinctions, indicating that they were completely uncorrelated in time.

Another issue was the notion of ecological equilibrium itself. In adaptational terms, individual species are always playing catch-up with environmental change, and they do so independently of one another. Over long periods of time there isn't anything resembling stasis because individual species come and go, just as do the individual ecosystems in which they resided. Mammoths, which enjoyed a tricontinental distribution, surely had broad

1. Thylacoleo or marsupial "lion" *(Thylacoleo carnifex)*
2. Thylacine or marsupial "tiger" *(Thylacinus cynocephalus)*
3. Red-necked wallaby *(Macropus rufogriseus)**
4. Mihirung *(Genyornis newtoni)*
5. Giant diprotodont *(Diprotodon optatum)*
6. Short-faced kangaroo *(Simosthenurus occidentalis)*
7. Tasmanian devil *(Sarcophilus harrisii laniarius)**
8. Wakefield's wombat *(Warendja wakefieldi)*
9. Ramsay's echidna or spiny anteater *(Megalibgwilia ramsayi)*
10. Hippolike diprotodont *(Zygomaturus trilobus)*

[* = an extant species]

PLATE 7.4. **NARACOORTE PANORAMA, SOUTH AUSTRALIA:** A national park and World Heritage site, the Naracoorte caves of southeastern South Australia have produced an enormous number of vertebrate fossils, ranging in age from 500,000 BP onward. Although a semiarid area today, in the Quaternary this region teemed with life. There are three carnivores in this scene: a thylacine bringing down a wallaby, a thylacoleo waiting in ambush on a eucalyptus limb, and a Tasmanian devil eyeing a wombat rooting about for a meal near a cave entrance. All are abundantly represented by fossils in the caves, indicating that this area was an excellent hunting ground. Also illustrated are short-faced kangaroos and two species of diprotodonts (plate 7.2). Although the Tasmanian devil is still extant, it died out on the Australian mainland less than 500 years ago.

PETER SCHOUTEN - 2008

PLATE 7.5 (PREVIOUS PAGE). **HERD OF WOOLLY MAMMOTHS ON THE SIBERIAN STEPPE:** Among mammoths, the matriarch-led herd was probably the fundamental social unit, as it is in living elephants. Herds would have been essentially family affairs, consisting of a mature female, her adult daughters, and their younger offspring. For much of the year, adult males would have presumably been solitary or part of temporary bachelor associations, interacting with females only when the latter indicated sexual receptivity. (In this scene, a male approaches the herd on the left, perhaps in a breeding state of mind.) At the onset of sexual maturity, male offspring would have been driven out of the herd. Suddenly forced to fend for themselves, they had to quickly figure out how to eat, take care of themselves, and avoid making poor decisions—not unlike human teenagers.

niche tolerances, even if they preferred steppe and grassland environments (see plate 7.5). It is hard to see why one kind of habitat, like the high-latitude mammoth steppe, could not have been exchanged for other habitats farther south, even if it meant range and population contraction. In the case of horses, how can there be any basis for claiming that the modern-day mustangs of Nevada, Utah, and other western states are somehow less adapted to exactly the same places in which their Pleistocene forebears lived? Their burgeoning populations suggest otherwise.

Martin, always interested in the big picture, rejected local explanations for Near Time losses based on climate change not because they were necessarily implausible on their face, but because there was nothing general enough that could be extracted from them to provide a basis for framing common-cause hypotheses. For climate change to be admitted as a direct or even an incidental cause of Late Pleistocene extinctions, he argued that three propositions had to be satisfied:

1. There has to be evidence of significant climate change in and around the various times and places where such extinctions occurred.
2. The change or changes had to have been unique within the late Quaternary, whether alone or in combination with other factors, to explain why the losses took place when they did rather than at any other time.
3. There must be a logical basis for the change or changes to have principally affected large terrestrial animals.

Of chief importance to Martin was the matter of degree: any transformation momentous enough to have caused substantial losses in one place had to have been of enormous magnitude, surely large enough to have had detectable impacts elsewhere, if not everywhere. This was not the perceived pattern, however: megafaunal extinctions in South America apparently occurred more or less in phase with those in North America, but nothing seems to have happened at the end of the Pleistocene in Antillea. As always, there was the inevitable question about size: why were larger animals expressly at risk? No prior extinction event in the Cenozoic exhibited this special feature, in which large size mattered in absolutely the wrong way.

For Martin, the failure of climate change supporters to adequately answer these points was a devastating critique, but others were not so sure. Requiring that unique, and possibly extinction-provoking, climatic oscillations had to be identified first was setting a very high bar, perhaps too high. The climate data available fifty years ago provided only highly compressed glimpses rather than intimately detailed pictures of what the past was like, and just because such oscillations were difficult to characterize did not mean they lacked biological impact. As noted in previous chapters, there are inherent physiological characteristics shared by large mammals, such as their relatively low reproduction rates, that could make them prone to extinction under many different circumstances. To turn Martin's criticism around, just because human predation may have marked certain large vertebrates for destruction in some contexts is not evidence that this happened in all contexts.

In sum, Martin was looking for an explanation for prehistoric losses that tied everything together, which he thought he had found in the idea of "dreadful syncopation"—the devastating, recurring blows to biodiversity that seem to have occurred in place after place, virtually worldwide, shortly after humans entered previously unvisited or unexploited environments. By contrast, the explanations developed by supporters of climate change were not covering arguments in the same sense, and from Martin's standpoint they were therefore always inadequate. For example, even if it were true that the elimination of woolly mammoths in high-latitude Eurasia was a result of increased seasonality and changes in plant distributions at the end of the Pleistocene, this event carries no logical implications for the disappearance of megafaunal lemurs in Madagascar 10,000 years later (see plate 7.6)—unless it could be shown that increased seasonality (or some ecological equivalent) applied to the latter as well. Martin considered that accounts of environmental flux and correlated loss were plausible enough within their specific geographical terms of reference, but they could not be generalized into a single, compelling set of cause-effect relationships that might be used to explain any comparable situation. Unlike overkill, he felt that their arguments lacked explanatory reach.

1. Grandidier's koala lemur *(Megaladapis grandidieri)*
2. Tretretretre sloth lemur *(Palaeopropithecus maximus)*
3. Long-armed sloth lemur *(Mesopropithecus dolichobrachion)*
4. Muller's elephant bird *(Mullerornis* sp.*)*

PLATE 7.6. **KOALA LEMUR AND SLOTH LEMURS IN ANKARANA FOREST, NORTHERN MADAGASCAR:** The common name for koala lemurs is based on the assumption that some of their locomotor habits might have been similar to those of the extant marsupial koala of Australia. The other large lemurs seen in this painting are sloth lemurs, whose elongated forelimbs were strikingly convergent on those of the true sloths of South America. The tretretretre sloth lemur (55 kg/120 lb) may have survived long enough to be recorded by a French author in the seventeenth century. Its relative, the long-armed sloth lemur, lacked the latter's conspicuous adaptations for under-branch suspension. It is more likely to have been a cautious "tree walker," as depicted here (see also plate 2.1). Many kinds of other native vertebrates inhabited the dry tropical forest that still covers the Ankarana limestone plateau, including Muller's elephant bird, a smaller version of the giant elephant bird seen in plate 1.1. Their remains are frequently encountered in the numerous caves and sinkholes that riddle the landscape.

8

OVERKILL NOW

PLATE 8.1. **A CUBAN CURSORIAL OWL** (*Ornimegalonyx oteroi*) swoops down on a young almiquí (*Solenodon cubanus*), an extant (and very highly endangered) native insectivore. The bird may have stood as much as 1 m (3.1 ft) tall, making it one of the largest, if not the largest, owl in the recent fossil record. Too big to fly, with highly reduced wings, it may have used its greatly elongated legs to literally run down or to parachute onto its prey, almost in the manner of an ambushing predatory mammal. Antillean cave deposits are often full of bones of small mammals and other vertebrates, dropped there by owls—the Quaternary paleontologist's best friends.

CURRENT STATE OF PLAY

Over the past five decades, Paul Martin's overkill hypothesis has excited debate in disciplines as diverse as archeology, paleontology, historical ecology, conservation biology, and even political economy and the philosophy of ideology. As we have seen, although a possible human role in prehistoric extinctions had long been suspected, until Martin's blitzkrieg hypothesis appeared, no one had ever claimed that human persecution was the only plausible cause for losses like those that swept through the continental Americas at the end of the Pleistocene. His was a shockingly radical departure from orthodoxy, and unsurprisingly it has engendered many an academic tiff, usually good-natured, between his detractors and supporters in scientific journals and at conferences.

However, these debates have not provided the clarity hoped for. The leading questions concerning the plausibility of overkill have remained essentially the same for half a century. Were humans always present immediately before the onset of major late Quaternary faunal collapses? If so, what were they actually doing to cause these losses and how fast did they occur? Why did islands suffer disproportionately? (See plate 8.1.) Why did the extinctions happen so quickly in some cases but not others? If large-bodied species were most at risk, how can numerous megafaunal survivals be explained, especially in Africa and southern Asia, where modern humans have lived for hundreds of thousands of years? And finally, how can we be so sure that climate change or other factors were not contributory in some way?

Thanks to continuing interest in the megafaunal extinctions, one beneficial outcome of the overkill controversy has been a steady increase in the worldwide pool of dates useful for accurately establishing when species disappeared. Radiocarbon dating methods also got better, and eventually cheaper, than they were when Martin first began exploring Near Time losses, and thanks to technical improvements in calibration and the introduction of mass spectrometry, smaller fossil samples could be used without sacrificing accuracy and precision. Cross-fertilization with other disciplines that have a potential bearing on extinction studies has also increased, in some cases exponentially. With the advent of ancient DNA techniques, questions about extinct species that were previously thought to be beyond any possibility of investigation have become progressively more tractable. Now past genetic diversity can be examined in a direct manner, permitting scientists to study whether collapsing diversity, genetic drift, hybridization, or other genomic phenomena like "invisible extirpation"—replacement of one population by another of the same species—might have played a cryptic role in some disappearances. As the following sections show, there have also been redefinitions or new definitions of what overkill might have entailed, with some authors suggesting the abandonment of the notion of blitzkrieg altogether.

RETHINKING CONTINENTAL LOSSES

New World

Quite apart from his speculations about causation, Martin's specific argument that North American losses were narrowly concentrated in time has held up surprisingly well, although it faces many other complications. Archeologist Stuart Fiedel has argued vigorously for the idea that the extinction window could have been as narrow as 400 years, during which much of the North American megafauna was either driven to immediate extinction or reduced to insupportably small numbers. In outline this is essentially the same picture that Martin originally painted of dramatic, rapid loss, but based on many more supporting dates. Other geographical areas have their own chronological and interpretative problems, but for North America, the locus classicus of Near Time extinctions, Martin was prescient regarding its speed, if not its cause (see plate 8.2).

A number of investigations conducted in the 1990s and early 2000s have also examined the second part of Martin's syncopation scenario, the supposedly short offset between the time of human arrival and the disappearance of local species. Among the most influential of these was a sophisticated simulation of overkill dynamics in end-Pleistocene North America published in 2001 by paleontologist and modeler John Alroy of Australia's MacQuarie University. In Alroy's view, the critical issue in testing overkill was to determine, in a probabilistic sense, whether human population growth could have been sufficiently rapid and hunting rates sufficiently high in early aboriginal North America to have caused the extinction of three-quarters of the continent's large herbivores between 13,500 and 12,500 years ago. To try to simulate conditions at the end of the Pleistocene, he defined a wide range of variables concerned with such factors as ecological dynamics, life history, meat consumption, and human population density. The trials started with 100 individuals as the minimum number of human immigrants entering the midcontinent 14,000 years ago, or just before the first indications of Clovis culture in the archeological record. In his preferred trial, the number of humans was allowed to grow to about 1,000,000 by 13,000 years ago, at a sustained annual growth rate of 1.66 percent (slightly higher than the rate used by Martin, but in the same ballpark). Massive losses are conceived to have begun just before maximum human population size was achieved, around 13,250 years go. The median time for extinction was 1200 years, with all extinctions completed by 1600 years after initial human arrival. As the extinctions wound down 12,400 years ago, human numbers stabilized. The model correctly predicted that there would be some megafaunal survivals in the aftermath of the period of major losses. Other runs produced the same overall picture.

For many observers, including Martin, Alroy's simulation study seemed to settle the climate-versus-people debate. However, if humans were already present in North America

1. Stag moose *(Cervalces scotti)*
2. Jefferson's ground sloth *(Megalonyx jeffersonii)*
3. Giant beaver *(Castoroides ohioensis)*
4. Jaguar *(Panthera onca)**
5. Long-nosed peccary *(Mylohyus nasutus)*
6. Harlan's musk ox *(Bootherium bombifrons)*
7. Vero tapir *(Tapirus veroensis)*

[* = an extant species]

PLATE 8.2. **APPALACHIAN PLATEAU PANORAMA:** The Appalachian Plateau is an elevated region in the eastern United States, running from southern New York to Alabama. At the end of the last ice age, its northern end would have been covered in boreal forest at higher elevations. Farther south, cool temperate forests like the one depicted here would have been the rule, with many watercourses and lakes. This was good habitat for giant beaver and stag moose, but also for an assortment of other large mammals—long-nosed peccary, vero tapir, Jefferson's ground sloth, and Harlan's musk ox. All of the species seen here are extinct except for the jaguar, which no longer lives in the United States except in isolated pockets along the Mexican border. This mixture of animals whose closest extant relatives live in areas ranging from the Arctic to the tropics may seem a bit bizarre, but it brings home the point that the faunal associations, as well as the diets, of Pleistocene species were very different from the ones we think of as "normal" today. The only megafaunal species confirmably living in northern Appalachia today are the black bear and white-tailed deer, although cougars lived there in the early twentieth century and may be returning.

1. Guanaco *(Lama guanicoe)*
2. Slender-headed mylodont sloth *(Scelidotherium leptocephalum)*
3. Cuvier's gomphothere *(Cuvieronius hyodon)*
4. Megatherium *(Megatherium tarijensis)*
5. Horse *(Equus* sp.*)*
6. Glyptodon *(Glyptodon* sp.*)*

PLATE 8.3. **PÁRAMO SCENE IN THE NORTHERN ANDES:** Páramo is the term for high-altitude alpine shrubland found in the northern Andes, below the perpetual snowline but above continuous forest at roughly 3000–4000 m (10,000–13,000 ft) elevation. Páramo environments are relatively cool but humid, experiencing high levels of solar radiation and daily temperature extremes of as much as 30°C (85°F). These conditions favor grassland and shrubland development, which is inviting to many kinds of mammals and birds. Today, the fauna includes spectacled bear, guanaco, coati, puma, tapir, and several rodent species, as well as condors, ducks, and several kinds of raptors that prey on the smaller mammals. In the Pleistocene, the fauna was even more diverse. Cuvier's gomphothere occupied many parts of South America, although it is best known from lowland contexts. Their presence in contexts dominated by páramo illustrates just how versatile, ecologically speaking, such large mammals obviously were. The same was true of horses and the giant ground sloths *Megatherium* and *Scelidotherium*. Remains of these huge sloths have been found as high as 4700 m (15,500 ft) in the Peruvian Andes, as well as in areas that would have been very close to the margin of the Patagonian ice cap 18,000 years ago.

at least 15,000 years ago, as we now know that they were, then the modeled rate of increase should have peaked well before the last appearance dates of most of the megafauna. More critically, some of Alroy's other runs suggest that, if parameters were changed only slightly—to reflect lower hunting pressure and thus a lower rate of population increase—then the simulated losses might not have occurred in the same way or perhaps at all.

As radiocarbon records have improved, a few well-substantiated examples of late survival of now extinct continental species have also been found, implying that perhaps losses were not always either rapid or simultaneous. From a theoretical perspective, this actually makes sense, especially if overhunting is seen as the only driver, because there should have been places where hunting pressure would have been negligible over long intervals. Perhaps surprisingly, one of these places may have been the very heart of Beringia: horses and mammoths apparently survived in the Alaskan interior until about 10,500 years ago, a millennium later than expected. Other examples include controversial evidence for the apparent persistence of several ground sloth and glyptodon species until the early or even middle Holocene in South America (see plate 8.3).

Speaking of sloths, an unusually interesting case concerns the native sloth species of Cuba, Hispaniola, and Puerto Rico, islands situated right next door to the continental Americas (see plate 8.4). Martin participated in a paper, as did I, that established on the basis of a host of new radiocarbon dates that several sloth species persisted for at least a millennium after the presumed arrival date of the first people about 6000 years ago. Martin wondered if the relatively small Antillean sloths might have escaped early elimination because they were rarely encountered, perhaps because the humans were focused on marine rather than terrestrial resources. Other work showed that the rest of Antillea had also been insulated from the faunal catastrophe visited on the New World continents at the end of the Pleistocene. What kind of natural calamity could have doomed the mainlands but spared the islands only a short distance away? It had to have been people, Martin reasoned; they just took longer to get to the islands. For Martin, this was the primary lesson of the Antillean extinction story, and it remains one of the most telling (if indirect) arguments for human complicity in New World extinctions.

Northern Eurasia

Eurasia was always a problem for Martin's overkill argument, and he knew it. Early hominins of one sort or another are now known to have been present in Europe proper at least 600,000 years ago (and as much as 1.8 million years ago farther east, in the Caucasus)—and they were hunting. If fully modern humans had already differentiated from their precursor lineages as much as 300,000 years ago, as new evidence from North Africa and central Europe indicates, then humans just like us have been part of the Old World scene for a really long time. A relevant Late Pleistocene record comes from the site of Sopochnaya Karga in the Yenisei basin of central Siberia, recently excavated by Vladimir Pitulko and Alexei Tikhonov

PLATE 8.4. **FOREST SCENE IN HISPANIOLA:** Monkeys related to those living today in South America lived on Cuba, Hispaniola, and Jamaica. Skeletal evidence for the Hispaniolan monkey (*Antillothrix bernensis*), seen here, suggests it was medium-sized (4 kg/8.8 lb) and not very acrobatic. This also seems to have been the case with the other monkeys. Among the native sloths of Antillea, locomotor habits varied. Smaller sloths, weighing as little as 5–10 kg (11–22 lb), were arboreal, although the much larger *Parocnus serus* (70 kg/155 lb), also depicted here, probably spent much of its time on the ground. Both monkeys and sloths managed to survive initial human colonization, which began around 6000 years ago, but all had probably disappeared before European colonization started in the sixteenth century.

1. Straight-tusked elephant *(Palaeoloxodon namadicus)*
2. Spotted hyena *(Crocuta crocuta)**
3. European ass *(Equus hydruntinus)*
4. Hartebeest *(Alcephalus buselaphus)**
5. Merck's rhinoceros *(Stephanorhinus hemitoechus)*

[* = an extant species]

PLATE 8.5. **AFRO-ASIAN CROSSROADS—THE LEVANT:** The Levant region of southwestern Asia has been a continental crossroads for 15 million years, a fact reflected in the mix of African and Asian species seen in this panorama. Hartebeest is nowadays completely restricted to Africa, but in the Late Pleistocene it ranged north of Sinai as well. The spotted hyena is also now exclusively African, but during the Middle Pleistocene it had a more extensive distribution, from Britain to central Asia (see fig. 5.4); by the end of the Pleistocene it had disappeared from Eurasia, for reasons unknown (perhaps climate change, or competition from wolves). The species of straight-tusked elephant seen here would have been two to three times the size of African elephants living today. There is no evidence that any species of Merck's rhino survived much past the Last Glacial Maximum. The fate of the European ass is more mysterious; it may have died out in the late Holocene, or it may have hybridized with other equids.

of the Russian Academy of Sciences. Humans butchered a mammoth there 45,000 years ago, some 35,500 years before *Mammuthus primigenius* breathed its last in exactly the same part of Eurasia. Does this mean that Siberian hunters were just slower to wreak havoc, or that these elephants held out longer because they had become aware of humans as predators? Or is it that they responded better—for a time—to climate change than did their North American cousins?

Martin's problem wasn't lack of evidence of megafaunal extinction, for many Eurasian species had disappeared over the long course of the Pleistocene. Instead, what was missing was an identifiable period of *concentrated* loss, parallel to the one that occurred in the continental New World. The handful that did take place at the very end of the Pleistocene, at essentially the same time as the much larger events in North and South America, allowed Martin to argue for a transcontinental common cause. However, the majority of Eurasian losses seem to have occurred earlier in the epoch, well outside the latest Pleistocene blitzkrieg bubble. For example, the cave bear (*Ursus spelaeus*) and one or more of the several species of Merck's rhino (*Stephanorhinus*; see plate 8.5) may have disappeared during the Last Glacial Maximum, well before the end-Pleistocene. The bear was a thoroughgoing herbivore; this may seem a strange adaptation to find in an ursid, but plant foods make up a large part of the diet of most extant bears, which are best thought of as opportunistic omni-

PLATES 8.6 AND 8.7. **VANISHED RHINOCEROSES:** Today, the word "rhino" conjures up images of hefty beasts with alarmingly sharp horns lumbering about in the tropical parts of Africa and Asia. But in Near Time rhinos ranged more widely, especially in Eurasia, where they occupied suitable habitat from Spain to the Philippines, from the Arctic to the equator. Despite their evident adaptive success, all mid- and high-latitude rhino species disappeared before the end of the Pleistocene, including the two pictured here. Steppe and woolly rhinos were supermegafaunal, weighing around 2000–3000 kg (4400–6600 lb); they evidently preferred open habitats, and were obviously cold-tolerant. Surprisingly, the closest living relative of the woolly rhino (*Coelodonta antiquitatis*, pictured here on the Siberian steppe) is the extremely endangered Sumatran rhino (*Dicerorhinus sumatrensis*), which weighs only about a quarter as much. Nowadays, due to human persecution, this fast-disappearing species is found only in tropical forests on Sumatra, Borneo, and peninsular Malaysia, whereas in the Late Pleistocene it ranged over much of mainland southeast Asia as far west as India. The steppe rhino (*Elasmotherium sibiricum*, seen here with the extant saiga antelope *Saiga tatarica* in central Asia) differed from other rhinos in a number of ways, the most obvious one being its single, notably massive horn. Some wags have recently taken to calling *Elasmotherium* the "Unicorn"—a most unlikely model, I think, for the sleek, horselike but cloven-hoofed creature depicted in medieval paintings and tapestries.

vores. Merck's rhino was just one of several Eurasian rhino species adapted to temperate and even polar latitudes (see plates 8.6 and 8.7). All of these species significantly overlapped with Pleistocene hominins, living in the same areas at the same times, but as usual there is little or no direct evidence concerning their disappearance. Did they die out because forage was unavailable at the height of the last ice age, and thus they were actual victims of climate change? Or were the bear and the rhinos in a long decline for other reasons, their ranges having shrunk to a few discontinuous pockets? Or should we just assume that humans had to have been involved, and be done with it? In the end, these are just more examples, among many others, to which extinction theories of any stripe may be casually affixed because none can be effectively contradicted.

In addition to Eurasian losses that might have occurred earlier than Martin would have predicted, there were also ones that occurred much later than expected, contrary to the dreadful syncopation formula. Although we now know that a small number of megafaunal species like the giant Irish deer (see plate 2.3) lasted into the early part of the Holocene, it is the truly protracted losses that are of special interest. The best known example of this is the persistence of woolly mammoths on Wrangel Island in the East Siberian Sea until roughly 4000 years ago (see box, page 83), but it is not unique. Woollies also survived until about 6000 years ago on St. Paul Island in the Pribilofs, a small group of islands situated near mainland Alaska in the southern Bering Sea (see figure 4.2). Complicating any simple scenario of mammoth extinction, mainland woollies seem to have survived about a millennium longer on the Arctic periphery of northern Asia than they did anywhere in continental North America. The emerging pattern in northern Eurasia suggests a patchwork of losses rather than a singular collapse, with mammoth populations declining toward extinction at different rates in different situations.

Sahul

One other place in which the debate over the primary cause of late Quaternary extinctions remains red-hot is Sahul, the name for the conjoint landmass formed by Australia and New Guinea when sea level was low (see figure 3.4). By any measure, this area suffered grievous devastation in the terminal part of the Pleistocene (see plates 8.8–8.10): 90 percent of megafaunal vertebrate species disappeared, leaving nothing behind among the mammals larger than the red kangaroo (*Osphranter rufus*, up to 90 kg or 200 lb). This is higher than the loss factor for North and South America, which together lost about 75 percent of their megafaunal mammals. What happened?

Recently, a research team presented evidence indicating that humans were living at the cave site of Madjedbebe in Australia's Northern Territory as early as 65,000 years ago. Not only is this date well beyond radiocarbon time, there aren't many sites in Australia that are even half this old. This may mean that that for many thousands of years humans

SPORORMIELLA, THE END-TIMES FUNGUS

Scientists like to have the support of multiple proxies for their interpretations, since this increases the likelihood that they are making correct inferences. In recent years, an unusual proxy has been gaining favor because its behavior seems to correlate well with major fluctuations in megafaunal population sizes. The basic idea is simple: *Sporormiella* is a fungus that grows only on mammalian feces. Therefore, the thinking goes, the more robust the population size, the more feces there should be—and the more *Sporormiella* spores in lake cores or cave sediments. In North America around the 12,000 to 13,000 year mark, spore counts in dated cores notably decline, often to background levels, then rise again in the early to late Holocene. Fairly consistent results of this type have been obtained in western and northeastern North America, Madagascar, and more recently in Australia, in good correlation with the much earlier losses that occurred there. One problem with using this proxy is that suitable coring sites are not necessarily optimally located with respect to likely megafaunal distributions or migration routes. Another is that the actual causes of dramatic oscillations in spore frequency still have to be inferred from other evidence: strong dips regionally may signify the onset of human persecution, or may relate to climate change or other agencies.

In addition to recovering spores and pollen, using a variety of techniques scientists also attempt to recover animal and plant DNA from sediments. In persistently cold contexts like northern Yukon, DNA survives reasonably well not only in bones and teeth, but also in the soil itself, in the form of short, isolated fragments. Short cores taken from frozen sediments can be analyzed in the lab to detect specific "fingerprint" sequences, which may in turn be used to determine species occurrence even in the absence of identifiable physical remains.

1. Thylacoleo or marsupial "lion" *(Thylacoleo carnifex)*
2. Marsupial "tapir" *(Palorchestes azael)*
3. Anak wallaby *(Protemnodon anak)*
4. Strong-beaked crocodile *(Quinkana fortirostrum)*
5. Paula's tree kangaroo *(Bohra paulae)*
6. Hippolike diprotodont *(Zygomaturus trilobus)*

PLATE 8.8. **FOREST PANORAMA, SOUTHEASTERN AUSTRALIA:** The forested regions of southeastern Australia are still among the most diverse on the continent, and, like elsewhere, their perennial waterways attract wildlife. In this scene, a marsupial "lion" attacks a pair of bizarre-looking marsupial "tapirs." At 200 kg (440 lbs), with probable semiaquatic habits as seen here, *Palorchestes* has sometimes been compared to placental hippos (but see plate 11.2). *Thylacoleo* weighed 100–130 kg (220–260 lb), which is the size of a large mountain lion. Its bolt-cutter jaws were able to exert an enormous amount of force relative to its body size, and it may well have been the ambush predator it is depicted as being here. Also seen is an extinct tree kangaroo, with pouch young, and a pair of large wallabies in the far distance. The strong-beaked crocodile on the right, a member of an extinct group of Pacific crocodylians (plate 12.2), is an immature animal. Adult members of this fully terrestrial species are estimated to have attained a length of up to 6 m—comparable to the goanna (plate 2.2), Australia's largest Pleistocene predator.

were not an especially common faunal element in Sahul. Significantly, between Madjedbebe times and these much later sites the record is objectively silent, giving no indication regarding how humans and megafauna interacted, if they did at all. With the exception of fossils from one controversial and possibly markedly disturbed locality, Cuddie Springs in New South Wales, there is nothing in the form of modified bones or the like that qualifies as direct evidence of human impacts on the Late Pleistocene fauna. So limited a database is of no practical value. Clearly, if there is an anthropogenic signal to be found, much more work will be needed to identify it.

On the basis of extensive excavations in Tasmania, archeologist Richard Cosgrove of La Trobe University in Melbourne has shown that 40,000 years ago people were hunting wombats and wallabies, but not the larger megafauna. Cosgrove thinks that this might have been a conscious decision on the hunters' part, because smaller mammals have a better fat-to-meat ratio than do larger ones—an important footnote to Martin's idea that it was all about the meat, and therefore the larger the prey the better. Cosgrove adds that it is also possible that the megafauna was already gone by this time, or so severely reduced in numbers that there was little or no hunting pressure brought to bear on them.

If it was not hunting pressure, what could have prompted the disappearance of almost all of the Sahulian vertebrate megafauna? Australia has been subjected to increasing aridification since the last part of the Pliocene, nearly three million years ago, but its trajectory toward desertification has by no means been constant. Indeed, the chief fact of life for Australia's biota has always been the continent's extreme climatic variability. As elsewhere, wetter/drier intervals alternated during the Pleistocene, in tune with atmospheric and oceanic thermohaline circulation patterns influenced by glaciation at mid- to high latitudes. However, in the western Pacific environmental effects associated with these intervals seem to have been greatly amplified (as happens on a smaller scale today with the phenomenon known as El Niño–Southern Oscillation). Beginning 40,000 to 50,000 years ago, Australia, and to a lesser extent New Guinea, experienced a period of truly intense aridification. The timing of this event is crucial, for it precisely matches the time during which the majority of megafaunal extinctions are now believed to have taken place. Inland Australia became exceptionally dry, with open, low-density dry forest being replaced by the kind of inedible sclerophyllous scrub (plants with small, leathery leaves) that covers much of the interior today. Aridification would have been less drastic along the eastern coast of Sahul at this time, especially in northern Queensland and neighboring New Guinea, thanks to the prevailing trade winds and annual monsoon. Conceivably, these areas might have acted as refugia for a time, but in the end it seems to have made no difference. The last proxy evidence of megafaunal presence—a ubiquitous fungus, *Sporormiella*, that thrives exclusively on mammal dung (see box, page 121)—declines precipitously in sediment cores dated to around 41,000 BP, while charcoal goes way up. Although humans were probably also disadvantaged by aridification, unlike other megafauna they

PLATE 8.9. **NINJA! A GIANT MEIOLANIID TURTLE** (*Ninjemys oweni*) and a short-faced kangaroo (*Procoptodon raphe*) forage contentedly in eucalyptus forest near Lake Eyre in south-central Australia. Not only did Australia specialize in huge lizards and snakes, it also possessed some of the largest turtles in the Pleistocene world. The body size of *Ninjemys* has not been adequately modeled, but it was probably around 200 kg (440 lb). Meiolaniid turtles, all now extinct, were once found in many places in the South Pacific, including New Caledonia, Lord Howe Island, and Vanuatu. *Ninjemys* probably died out around 40,000 BP in Sahul, but at least one related species managed to survive on an island in Vanuatu until about 3000 years ago—just when humans arrived. Its butchered bones in archeological sites tell the tale. On a pleasanter note, AMNH paleontologist Eugene Gaffney thought it would be imaginative—not to say hilarious—to salute the Teenage Mutant Ninja Turtles by naming this genus after them. Now that's totally tubular!

managed to endure. Yet there is nothing to indicate that humans delivered a blow of any sort, not even a coup de grâce, to the waning populations of large mammals, birds, and reptiles (see plate 8.9).

For reasons like these, few Australian archeologists and paleontologists support over-hunting at present, most preferring instead to view environmental change as the extinction driver. However, Tim Flannery, a prominent Australian mammalogist, places a very different spin on the few facts available. His argument implicates humans not only in causing the extinctions but also in transforming Australian environments. In this variant of the usual blitzkrieg scenario, it is fire, in human hands, that is the agent of destruction. Flannery allows that overhunting as such also took place, but in his view the kill mechanism is better conceived as being environmental change provoked by early humans habitually burning the fuel-laden, tinder-dry plant cover to flush game or facilitate simple agriculture. By 40,000 to 45,000 years ago, dry broadleaf forest, formerly dominant over much of the eastern part of the continent and already stressed by desertification, would have been subjected to an unnatural fire regime, either because humans were overburning or because with the loss of browsing megafauna, tinder had built up to dangerous levels on forest floors—or perhaps both factors. Unable to recover from increased levels of catastrophic incineration, the forest retreated to pockets along the eastern periphery of the continent. Sclerophylls, armed with strong chemical defenses against browsing mammals, expanded throughout the interior of Australia, creating the biomes seen today. New Guinea was less affected, but ultimately suffered in the same manner (see plate 8.10). Trapped between deleterious environmental changes and human persecution, the megafauna had nowhere to go and simply collapsed. While this scenario differs from classic blitzkrieg, humans still stand at the head of a causal chain of events that resulted in numerous losses, which in the logic of the extinctions debate makes them ultimately culpable.

In another version of the overkill theme, modeler Frédérik Saltré and colleagues vigorously challenged the idea that climate change could have been responsible for Sahul-

PLATE 8.10. **MARSUPIAL "PANDA," NEW GUINEA:** The fossil record of the New Guinea portion of Sahul is more limited than that of Australia. The fauna was apparently less diverse, especially in regard to large-bodied species and ones adapted to fairly dry conditions. *Hulitherium tomasetti*, seen here, was only about 100 kg (220 lb), much smaller than its Australian hippo-sized relative, the diprotodont *Zygomaturus* (plate 8.8). *Hulitherium*'s stumpy proportions are vaguely like those of the panda, and like the latter it was a browser. On the branch above is the ribbon-tailed bird of paradise, *Astrapia mayeri*, an extant species.

ian losses. Their simulations indicated that the Australian extinctions were uncorrelated with climatic variability over the last 120,000 years, shifting, at least rhetorically, the blame back toward anthropogenic effects. However, one of their necessary assumptions was that humans and fauna overlapped for 13,500 years, already a very long interval for classic overkill to run to completion. Now that there is some evidence that humans were already in northern Australia 65,000 years ago, the period of overlap must be extended to something like 23,000 years. Although the authors argued that in the right circumstances ordinary hunter-gatherer levels of exploitation can result in complete extinctions, an anthropogenic cause that extends over almost half of Near Time—with no acknowledged assistance from climate and without any species recovery—is very different from anything Martin conceived.

All sides agree that the thing which chiefly prevents further progress in settling the cause of Sahulian losses is the continuing paucity of good dates. Accurate and plentiful dating is an essential form of evidence for understanding any extinction event, and for Australia and New Guinea this is not a problem that will be easily resolved. A couple of hundred Late Pleistocene localities have been identified in Australia, but fewer than two dozen have yielded dating results within radiocarbon time. There are good reasons for this. Probably many of these sites are in fact well beyond the limits of radiocarbon time—i.e., they are older than 50,000 BP. There is also the problem of sample degradation. Protein (often called collagen, after the most abundant, but not the only, type of protein present in bone) is the material generally used for directly dating bones and teeth, but it does not survive well in hot conditions like those associated with shallow caves and ancient lake bottoms, the sorts of contexts from which the majority of Australian Pleistocene fossils have been recovered. OSL dating has helped to fill in some gaps, and indeed provides the strongest available evidence for the proposition that the great preponderance of megafaunal extinctions happened a bit more than 40,000 years ago. The court of scientific opinion is still out, but if a majority of the losses really did occur 10,000 to 20,000 years after human arrival, classic overkill as an explanation is in trouble.

Africa

Martin estimated that about 15 percent of the Late Pleistocene African megafauna died out, as compared to about 75 percent in North America—a huge difference that demanded explanation (see plate 8.11). He argued that Africa managed to escape experiencing a Late Pleistocene faunal crisis because, unlike the New World, megafaunal species and anatomically modern humans had experienced each other for tens to hundreds of thousands of years. As a result, prey species were able to keep up with humans constantly improving their tool kits and hunting strategies by learning new avoidance behaviors and passing them on to offspring. In evolutionary terms, this is known as behavioral coevolution. Selection still went on—the animals that didn't learn fast enough still died, leaving the more perceptive among the hunted to contribute their genes to future generations. Also, Africa abounds in predators, and prehistoric human groups may have been under pressure from animal competitors. This might have been beneficial, enforcing equilibria between prey and their many predators. Human hunting occurred, of course, but sustainably, far below levels that might force extinction.

Thanks to continuing archeological investigations, evidence of hunting by anatomically modern humans during the latest Pleistocene and Holocene of Africa has greatly increased over the past fifty years. Almost all of it concerns species that are still extant, with very little indication that Near Time humans went determinedly after species that are now extinct. Coordinated major losses of the sort specified by the overkill scenario just did not happen.

Of course, other interpretations for the lack of concentrated losses in Africa are possible, and researchers have wondered why, quite apart from Martin's notion of behavioral coevolution, the Late Pleistocene pattern of loss on this continent was so different from elsewhere. Was it the "safety valve" represented by consistently available, extensive grassland habitat that enabled most of the herbivorous African megafauna to survive the vicissitudes of Near Time? During the most recent phase of intense aridification in Africa, which lasted from about 26,000 BP to 16,000 BP in partial correlation with Northern Hemisphere glaciation, grasslands expanded into regions formerly occupied by rain forest, to the great benefit of those same herbivores. As conditions became warmer and moister through late glacial time and into the early Holocene, equatorial rain forest returned, but by this time grassland was beginning to extend into the Sahara because the west African monsoon became more intense. For at least some of the megafauna, the period between 4000 and 7000 years ago must have been the time during which they achieved their maximum distributions. After the mid-Holocene, the grasslands of Africa retreated to their modern positions. But the megafauna had mostly survived, whatever the level of human persecution.

RETHINKING INSULAR LOSSES

In sharp contrast to continuing disagreement about the cause of continental extinctions, both blitzkriegers and climate change enthusiasts are in agreement that very recent island losses—those that occurred within the modern era, or the last five hundred years—were driven by human activities. Thus in the case of Antillea, a wave of losses occurred around the time of European discovery and early occupation, affecting most of the remaining mammals and a number of birds. These extinctions were presumably caused by environmental degradation, land clearance, and the introduction of invasive species (especially rats, mon-

1. Gorgon hippopotamus *(Hippopotamus gorgops)*
2. Plains zebra *(Equus quagga)**
3. Reck's straight-tusked elephant *(Palaeoloxodon recki)*
4. Cattle egret *(Bubulcus ibis)**
5. Juma giraffe *(Giraffa jumae)*
6. Giant narrow-horned hartebeest *(Parmularius angusticornis)*
7. Sivathere *(Sivatherium maurusium)*

[* = an extant species]

PLATE 8.11. **EAST AFRICAN RIFT PANORAMA:** For millions of years, the fertile volcanic soils of the East African Rift have supported many kinds of herbivorous mammals. These sediments are also excellent for fossil preservation, and as a result we have an unequaled record of Pleistocene animal life for this part of Africa. Most localities, however, are Early to Middle Pleistocene (as is the scene depicted here). The gorgon hippo, giant hartebeest, and zebra are similar to their living relatives. With its long neck and legs, the Juma giraffe closely resembles the extant African giraffe, but the sivathere—although also a giraffid—looks very different. The zebras provide a body size scale for Reck's elephant, a relative of the modern forest elephant but very much larger. The extant cattle egret, which now enjoys a nearly worldwide distribution in tropical to temperate areas, probably originated in Africa or adjacent parts of Eurasia. Its fate has been the reverse of that of the megafauna: association with humans and their domestic animals has led to amazing success for this species, rather than extinction.

PLATE 8.12. **JAVANESE DWARF STEGODON AND WATER BUFFALO:** The landmasses that now comprise most of western and central Indonesia were intermittently connected to mainland southeast Asia during periods of low sea level. This permitted mainland species to easily migrate into this area, known as Sundaland (see figure 3.4). Stegodons, related to living elephants but constituting a separate family, were a characteristic part of the Pleistocene fauna of many Indonesian islands. As in the case of elephants migrating to islands in the Mediterranean (plate 10.3), miniaturization was the usual consequence. The Javanese species *Stegodon hypsilophus* is mostly known from the Middle Pleistocene, but elsewhere the genus persisted until the later Pleistocene. The extinct water buffalo *Bubalus palaeokerabau* does not differ greatly from the extant "wild" Asian water buffalo (*Bubalus arnee*), and may therefore not be a distinct species. In general, although mammalian and avian extinctions occurred on these islands during the Pleistocene, losses were neither extensive nor concentrated until the arrival of anatomically modern people.

gooses, and cats) over the course of several centuries. Recent bird extinctions throughout the Pacific region were doubtless driven by similar kinds of impacts.

But not all island losses are very recent, and new work has yielded a number of surprises. The big news comes from the dating of multiple vertebrate extinctions in a part of the habitable world for which Martin had very little data fifty years ago—the island groups east of the Asian mainland, from Indonesia to Philippines to Japan. Important discoveries have been made in recent years on many of them, but the ones of special interest come from the island of Flores in eastern Indonesia, famous for the previously mentioned discovery of the extinct hominin *Homo floresiensis* (see chapter 4). At the site of Liang Bua, paleontologists have identified a number of nonhominin megafauna, including giant scavenging storks and dwarf stegodons (see plate 8.12). These species persisted until about fifty thousand years ago; thereafter, nothing. This of course sounds like the usual island extinction story, save in one important regard—among the deaths, there was one that occurred within the family. *Homo floresiensis*, which had inhabited Flores for hundreds of thousands of years, apparently without causing any extinctions itself, disappeared once *Homo sapiens* arrived. . . .

9

WHERE ARE THE BODIES, AND OTHER OBJECTIONS TO OVERKILL

PLATE 9.1. **HORSES ENTERED SOUTH AMERICA** 2 to 3 million years ago during the Great American Biotic Interchange, when species of many different kinds migrated in both directions over the newly completed Panamanian isthmus. Horses from North America prospered in grassland regions like the llanos and pampas, and their fossils are common in South American Pleistocene sites from northern Venezuela to southern Patagonia. One of the first sites to produce the remains of extinct South American megafauna in clear association with human artifacts was Fell's Cave in southern Chile, dug by AMNH curator of archeology Junius Bird in the 1930s. He was initially worried that "if these bones should prove to belong to an animal introduced by Europeans, all our conclusions on our previous work were wrong." Fortunately, later studies—eventually involving radiocarbon dating—"proved that we had found the first evidence that this ancient horse was hunted and eaten by the early natives of South America."

WHERE ARE THE BODIES?

When Paul Martin started his investigations, available radiocarbon chronologies appeared to support to some degree the correlation between human arrival and sudden faunal disappearances, but other forms of evidence did not, or were neutral indicators at best. An important example of incompliant evidence concerns the almost universal lack of examples of mass kill sites in places where large losses occurred, such as Australia and the Americas. Such sites would be expected to contain not only the remains of numerous animals, but also signs of direct impacts in the form of butchered bones, discarded tools, processing stations, and other earmarks of human presence. If the extinctions occurred as swiftly as the blitzkrieg scenario required, theoretically such sites ought to exist in abundance. But they don't.

There is of course some evidence of megafaunal hunting—almost always of one or a few animals at a time—in early aboriginal North America. Most examples concern bison and the two proboscideans I introduced in chapter 1 (the mammoth *Mammuthus* and the mastodon *Mammut*). For most other North American megafaunal species, there is no indication of fatal interactions with humans at all, or at least nothing that's very convincing. And absolutely missing is anything resembling a prehistoric shambles with megafaunal remains piled in gruesome windrows, tracking the spread of humans across continents in the way that Martin envisaged. But Martin had an explanation for the apparent paradox: "An explosive model will account for the scarcity of extinct animals associated with Paleo-Indian artifacts in obvious kill sites. The big game hunters achieved high population density only during those few years when their prey was abundant. Elaborate drives or traps were unnecessary."

The debate over what all of this means remains evergreen and predictable. Overkill supporters say that the lack of mass kill sites doesn't matter very much, because fossils are always rare, and, anyway, absence of evidence is not evidence of absence. Anti-overkill supporters wonder how this could possibly be the case if *over*hunting was what actually took the animals down. Killing at a prodigious scale implies that a lot of indicators of human activities should have been left on the landscape, and at least some of these ought to have good representation in the archeological record. I'm not talking about the occasional Clovis point stuck in a vertebra, or a bone with cut marks here and there. I'm talking about evidence of *intensive* exploitation, whatever that might have consisted of during the overhunting period. Yet even after many decades of archeological investigation, the fact remains that a megafaunal mass kill site has never been identified in the Late Pleistocene of North America. The only contexts that even come close are the buffalo jumps found in the northern plains of the United States and Canada, where herds of bison were purposely stampeded over cliffs. Although some jumps might have been utilized in the latest Pleistocene, almost all of the well-dated ones are much younger than 5000 BP. In any case, whatever the impact

PLATE 9.2. **BISON ENTERED NORTH AMERICA FROM ASIA,** where they first evolved, over an earlier version of the Bering land bridge about 180,000 years ago. These newcomers did very well in North America, rapidly extending their range throughout the continent. The species *Bison latifrons* ("broad forehead"), depicted here, was so named for a fairly obvious reason—those immense horns. The spread between horn cores (horn sheaths do not usually preserve as fossils) may have been as much as 2.4 m (8 ft), according to some estimates. But in other respects *B. latifrons* seems to have just been a very large, modern-looking bison. We are not sure how distinct it was genetically, as efforts to collect its DNA have so far been unsuccessful. In the painting, the forequarters are depicted as not being especially hairy, a difference from living *B. bison*. This accords with the observation, made on the basis of living bison, that animals with large horns tend to exhibit less "display hair."

PLATE 9.3. **THE MORNINGSTAR-TAILED GLYPTODON** *(Doedicurus clavicudatus)* was among the last of the giant armadillos. A massive animal, it weighed as much as 2000–2400 kg (4400–5275 lb)—by comparison, the largest living armadillo weighs only about 25 kg (55 lb). All Pleistocene glyptodons possessed armored tails, but in *Doedicurus* the tube had evolved into a formidable club, tipped by clusters of conical spikes. The tails may have been used to drive off potential predators, but their real function would have been in intraspecific combat, to achieve dominance or access to females. In this scene two males are squaring off: the one in the foreground has just smashed the business end of his tail against the carapace of the other, who is evidently clearing his mind before making his next move. That such encounters actually occurred is supported by the discovery of carapaces with deep indentations, acquired in life and presumably the result of interactions like this one.

this form of wasteful hunting may have had on prehistoric bison populations, the long-term effect was nugatory: *Bison bison* is a survivor.

In South America the picture is similar. Among the few megamammals found in confirmed or tolerably likely kill sites are the gomphothere *Notiomastodon*, the glyptodon *Doedicurus*, the giant sloth *Megatherium*, and horses (see plates 9.1, 9.3, and 11.3). Although the range of megafaunal targets is slightly greater than that for North America, guanacos and deer are by far the most frequently encountered large mammals in archeological sites of this age. These too are survivors.

Gary Haynes, a professor of archeology at the University of Nevada at Reno and a strong proponent of overkill, has given the problem of absent bodies a good deal of thought, particularly in regard to the Clovis archeological culture (in Haynes's opinion, the chief if not the only perpetrator behind North American overkill). Well-dated Clovis sites range in age from 13,400 to 12,900 years ago. He agrees that indisputable associations between remains of megafaunal animals and Clovis cultural deposits are quite uncommon, but he believes that this is because they are both rarely encountered and rarely recognized for what they are. He also cites modern evidence to the effect that kill site refuse, in the form of bones and teeth, tends to disappear quickly, thanks not only to the action of the elements but also to scavenging carnivores, invertebrates, and microbes. Although he concludes from actualistic studies of recent bone deposition that preservation of skeletal remains is always statistically unlikely, both in cultural and noncultural contexts, such studies do not begin to cover the same time range, or the same variety of depositional circumstances, as fossil-bearing areas of Late Pleistocene age. The problem of exceptional rarity remains.

Yet supporters of overhunting do have a point. To appreciate this, compare the taphonomy of Late Pleistocene extinction in North America to the K/Pg Mass Extinction at the end of the Cretaceous 66 million years ago, which resulted in the disappearance of all nonavian dinosaurs and possibly as much as 75 percent of all species then living. Remarkably, the western interior of North America is one of the very few regions on the planet that has yielded numerous dinosaur fossils that actually date to precisely 66.02 million years ago, when the Chicxulub impactor struck. We assume that nonavian dinosaurs living elsewhere were driven to extinction at the same time as the tyrannosaurs, duckbills, and ceratopsids of North America, but the primary support for this speculation is negative evidence: there's simply no indication, anywhere, that any dinosaurs other than birds made it past the Mesozoic–Cenozoic boundary.

With regard to Near Time losses in North America, last appearance dates for a number of megafauna correspond well with those for mammoths and horses. However, the number of good dating records is far fewer for these other species, and because of that it is hard to tell whether they disappeared quickly or slowly. Moreover, just because a terminal date falls within the period of human residency does not automatically become evidence that the species in question was hunted to extinction. For example, extinct species of camels were common in the Late Pleistocene of western North America, but there is only a single locality that has yielded solid evidence of actual camel hunting, a site 12,000 years old named Wally's Beach in southern Alberta. Can we base anything at all about human involvement in the disappearance of camels on so slim a foundation? It all depends on how one chooses to treat single observations, as either representative of a much larger constellation of similar observations just waiting to be made, or as uncorroborated and thus uninformative (or minimally informative) until further notice. Although there is something to be said for both sides of this argument, in my view propositions based on an absence of evidence always need to be treated with a fair amount of caution.

Haynes concentrated on proven Clovis localities, which was reasonable from his standpoint because he believed that Clovis culture was not only the first but also the only big-game-hunting tradition in North America at the time of the megafaunal extinctions. The sixteen North American sites for which there is substantial evidence of mammoth or mastodon hunting have an age range of about 12,000 to 14,500 BP. That works out to an average of one kill site every 150 years, which does not seem like much of a blitz even if this sample underestimates the true number of sites by a couple of orders of magnitude. Definite Clovis artifacts are unknown (or have not been recognized as such) for the first thousand years of this range. Were hunters from a different tradition interacting with these huge mammals for at least a millennium before Clovis people came on the scene? This possibility now seems quite likely. The Page-Ladson mastodon site, mentioned in chapter 4, dates to approximately 14,500 years ago and contains artifacts that are definitely non-Clovis according to its excavators. This need not mean that Clovis people were completely absent from the stage this long ago, as opposed to undetected, but it does mean that dreadful syncopation—Martin's cornerstone argument—becomes increasingly harder to support if several human groups had already passed into North America and were interacting with the megafauna by or before 15,000 BP.

PALEOECONOMICS OF OVERHUNTING

Words always come with connotations, whether these pop into one's conscious mind when a term is heard or read, or travel beneath the surface as subliminal memes. So it is with "overhunting." What exactly are the implications of this heavily freighted word? If humans were conducting themselves in the continental Americas at the end of the Pleistocene in the way specified by Martin's hypothesis, were they going after game animals in a way that could be reasonably identified as "hunting," as opposed to "casual slaughter," "unquenchable bloodlust," or, indeed, "overkilling"? In other words, were the hunters performing a recognizable, purposeful task, one that presumably required some level of training, planning, and forethought, and were they doing it for some equally recognizable reason? After all, surely we have no basis for believing that these hunters went about their business *intending* to cause extinction, as though that were the actual goal they had in mind. Of course, Martin considered that the likeliest purpose of overhunting was to secure continuous supplies of meat to ensure high birth rates, no matter how wastefully conducted their acquisition was in practice. But he never really explored the ramifications of this assertion, with its semi-Malthusian determinism that more food must always mean more babies. In his later work, he de-emphasized meat-banking but did not abandon it.

In terms of a modern-world economic analog, hunting for bush meat for local consumption might be thought of as an activity comparable to Pleistocene overhunting, but in most respects it really isn't. In Africa, this kind of market hunting tends to be serendipitous rather than highly organized, and, most important for present purposes, its targets are usually easily acquired birds or small mammals such as bats, rodents, and monkeys rather than megafauna. For obvious reasons, hunting large mammals with the simple equipment used by market hunters would be to invite considerable, even mortal, risk for an incommensurate return unless it were routinely possible to completely utilize the kill. Bush-meat markets are generally not set up for that level of processing, let alone distribution.

A more appropriate analog is rhino and elephant poaching, which is conducted at scale and with a narrow focus. As deplorable as poaching is, at base it is an economic activity, and it needs to be interpreted as such for it to make sense. Despite current antipoaching legislation and monitoring, illegal takings of elephant and rhino could well continue into the future, especially if prices rise as animals become rarer. Yet it is also reasonable to assume that at some point, for whatever reason, poaching will cease or at least diminish considerably as a market-driven activity. This may happen because tastes change, or because the cost of poaching becomes higher than the market can sustain, or the animals become too hard to find. If this last comes to pass, it will be an open question whether populations of targeted species will be large enough to permit their recovery.

Martin's prehistoric overkill model has similar features: targets were preferentially selected (large size), specialized equipment for big-game hunting was required (weapons, other utensils), and, at least to begin with, relatively small numbers of people were involved (see plate 9.4). Like poachers today, early big-game hunters in the continental Americas could have been immensely profligate, taking from their kills only what they immediately needed and leaving the rest, as there were always more animals to bring down. Although the global archeological record indicates that tusks and bone were occasionally collected for various purposes, especially toolmaking and adornment, acquisition of useful raw materials was not the primary reason for overhunting in Martin's conception.

However, here it is important to note that, having made a kill, the paleohunter hasn't actually secured his meat supply for its intended social purpose. This happens

PLATE 9.4. **HUNTING MERCK'S RHINO** *(Stephanorhinus kirchbergensis)*: Although this scene is set in southern Eurasia and the hominins are Neandertals rather than fully modern humans, the question is the same: Could large-mammal hunting by humans with simple tool kits have ever been conducted at the level necessary to cause widespread population collapses?

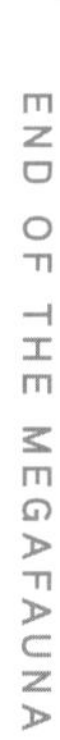

only when the meat is deposited at a place or places where it can be distributed and consumed, presumably by his kin group, which brings up the problem of transportation and perhaps storage. Selecting choice cuts and leaving the rest behind is arguably an ancient practice: the tongue of the mammoth slaughtered at the northern Siberian site of Sopochnaya Karga 45,000 years ago was removed by slicing away its attachments to the hyoid apparatus (a series of small bones under the jaw that support tongue muscles), while certain other parts, perhaps equally accessible, were evidently not removed. Obviously, if the goods had to be man-hauled, a tongue, which is solid muscle and easily butchered, has advantages over a joint that requires much preparation. Some other late Quaternary kill sites indicate apparent selectivity in what the hunters carefully processed or purposely left behind.

Yet, in the end, parallels between poaching and prehistoric overhunting only go so far. Modern poachers possess many advantages compared to paleohunters. They rely on rapid transportation, in the form of vehicles and perhaps helicopters or light planes in some places. Compared to hunters in Late Pleistocene North America, they have effectively infinite numbers of projectiles (bullets) to spend, which require no effort to launch apart from pulling a trigger. Experience or finesse in making a clean kill is unimportant to the well-armed poacher, whereas hunting a glyptodon probably posed tactical problems for ancient hunters in South America that were quite different from those involved in hunting a gomphothere, a ground sloth, or a meridiungulate (see plate 9.5).

Finally, it seems unlikely that megafaunal species living in the continental New World at the end of the Pleistocene would have reacted to human hunters in a uniform manner, whatever their level of naïveté. Some species should have been easy targets in many circumstances, others less so or not at all. Others should have persisted in areas that the humans did not or could not immediately penetrate. In earlier chapters I have pointed to a handful of examples of late survival in places where major losses occurred, but that's the point: there aren't many. The radiocarbon record, to the degree that it may be considered representative, implies that New World losses were sudden rather than staggered. Evidently, when the extinction event was over, it was *over*, except for a few hopeless stragglers. But was this due to Martin's overhunting humans surging across the continental Americas like a terrestrial tsunami, taking out nearly everything large in their path, or to some other factor or factors?

Bands and Surpluses

For it to be effective as a cause of extinction, overkilling requires enormous levels of slaughter within narrowly confined periods of time, with no chance of species recovery. Assuming that overkilling actually occurred in end-Pleistocene North and South America, the paleohunters must have produced many more resources in the form of carcasses than they could have possibly utilized at any given moment. In short, they created surpluses, at least potentially, even if they never used them. This raises several interesting points about overkill, some of which seriously conflict with inferences often made about the organizational capacities of prehistoric cultures.

Archeologist George C. Frison argued that the structure of the smallest human societies, formally denoted as bands, do not have the kind of social organization or cohesiveness needed to conduct extreme hunting or efficiently utilize surpluses. Certainly, in the North American archeological record for the end-Pleistocene there is no indication of anything

other than band-level hunting and gathering as primary economic activities. In Frison's view, human predation on mammoths and other megafauna certainly occurred, and may have had an appreciable impact in some instances, but it is more likely that other factors controlled which species survived and which didn't. Thus there is very little archeological evidence that horses and camels were hunted in North America, but they died out anyway; bison were hunted heavily, but survived.

As to meat caching, nothing in the record indicates that very large surpluses were acquired and stored by the first humans in the New World. However, paleontologist Dan Fisher of the University of Michigan thinks that the inventiveness of Pleistocene hunters has been undervalued. To illustrate this, he has undertaken some fascinating real-time experiments in storing meat in a manner that might have been available to early humans living in temperate regions. In one experiment, Fisher placed large pieces of meat in a

PLATE 9.5. **HERD OF *MACRAUCHENIA* AT A WATER HOLE:** Charles Darwin recovered the first fossils of this extraordinary horse-sized mammal in 1834 in southern Patagonia, during the *Beagle* expedition. He had absolutely no idea what it was. In recent years, molecular investigations have finally established that *Macrauchenia patachonica* is actually a distant relative of the group that today includes horses, rhinos, and tapirs (Perissodactyla). Its skull suggests that its face bore a large, fleshy proboscis of uncertain size and function. *Macrauchenia* has been found in many parts of South America, particularly areas that would have been dry grassland in the Late Pleistocene. Although it persisted until the beginning of the Holocene, there is no evidence that this species was ever hunted. By contrast, Darwin's rhea (*Rhea pennata*), a large flightless bird, was a frequent paleohunter target. Yet this species survived, and is one of the commonest sights in Patagonia today.

shallow lake that thoroughly froze during the winter, then collected the pieces in the spring for consumption by himself and his family. Despite what you are thinking, nobody got sick! Apart from a frothy coating of bacteria and small organisms wriggling about on external surfaces directly exposed to the water, which were easily removed, the meat had remained internally fresh enough for consumption without ill effects. Of course, whether anything resembling this method of preservation could have been widely practiced in end-Pleistocene North America can be debated. However, human ingenuity being what it has always been, other methods of storing leftovers (e.g., smoking) could have been utilized, and doubtless were. But then why continue to overhunt if a basic level of food supply could be achieved with minimal effort?

All of this raises the question whether early migrants into North America, presumably consisting of small, mobile kin-groups, could possibly have forced the extinction of dozens of species over areas amounting to hundreds of thousands of square kilometers merely by continuing to act in classic hunter-gatherer mode. To accomplish anything like the kill level required, these migrants would have had to have been organized into hunting parties on at least a semipermanent basis, and continue as such for long lengths of time. This flies in the face of ethnographic tenets which hold that bands organize into large groups for single, limited-time purposes, such as annual hunting initiatives, then dissolve when the task has been accomplished. Typically, the cooperation among individuals needed to achieve an immediate goal tends to end as soon as the goal has been attained. Further, given that it is characteristic of band-level societies to have weak, transient leadership, how would permanent hunting parties maintain their focus and discipline? What, if anything, was done with the mountains of meat and body parts in societies that characteristically do not acquire surpluses? If they fueled the end-Pleistocene human population explosion that Martin hypothesized, why did this not lead to a higher level of economic organization more quickly than is apparent in the archeological record? Or were the kills just abandoned, because the urge to overkill has nothing to do with rational economic behavior, but is instead something embedded in the genes of predators? That is a thought-provoking—even disturbing—question, and leads us to consider another, darker kind of surplus acquisition.

Surplus Killing—The "Henhouse Syndrome"

Killing prey over and above what would be needed for immediate food requirements is widespread among predators at every trophic level, from predatory zooplankton to large mammalian carnivores. Known as henhouse syndrome or surplus killing, the concept was originally proposed by zoologist Hans Kruuk in a 1972 landmark study of spotted hyena and red foxes. Kruuk argued that:

> [s]atiation in carnivores does not inhibit further *catching* and *killing*, but it probably does inhibit *searching* and *hunting* [emphasis in original]. Thus carnivores are able to procure

> an "easy prey" but normally satiation limits numbers killed. . . . Many, if not all carnivores possess behaviour patterns which allow utilization of a kill at a later time, or allow other members of the same social unit or offspring to use the food.

Since Kruuk's original work, a number of additional investigations of surplus killing have been conducted that tend to support and extend his ideas. For example, a study on introduced predators (dingoes and foxes) in Australia noted that "surplus killing events appeared to reflect ineffective anti-predator defenses by prey species when encountering a novel and efficient predator to which they have had no evolutionary exposure." Dingoes were introduced 5000 to 6000 years ago, foxes only a century and a half ago; both have had serious impacts on Australian wildlife. In the observed surplus killing events, dingoes tended to concentrate on the young of stock animals, perhaps because the latter are docile by nature and therefore easy marks (see chapter 10). Foxes are much more indiscriminate, killing anything within their preferred prey size range.

Kruuk also argued that, among predators, a possible evolutionary function of excess hunting is that it creates accessible surpluses for their kin—in other words, it is a form of social behavior that helps to ensure the continuance of the predator's genetic endowment. Allowing that earlier humans might have acted like other predators in this regard and killed beyond their immediate needs, a rationale for overkilling could be that it provided, at least in principle, surpluses that could be utilized by the hunters themselves or their relatives later on. But can overkill really be regarded as just an amped-up version of surplus killing? The complexities of human behavior make it difficult to reach a conclusion. Martin did briefly consider surplus killing in a paper cowritten with David Steadman of the University of Florida, but their conclusion was that it had little relevance for explaining Quaternary extinctions except, possibly, for some island losses—to which we shall now turn.

10

MORE OBJECTIONS: BETRAYAL FROM WITHIN?

PLATE 10.1. **GREAT AND SMALL, MOA AND WREN:** This image encapsulates, as well as any single image can, what has happened to island species everywhere during the late Quaternary: whether they were large or small, most have disappeared. The small birds are native New Zealand wrens, *Pachyplichas yaldwyni*, while the giant foot striding over them belongs to a female *Dinornis novaezelandiae*, the largest of the moa. The wren, probably largely terrestrial despite its small size, weighed no more than a few grams; the moa may have weighed as much as 250 kg (550 lb). Different species of wrens lived on the North and South Islands, and both seem to have survived into the Polynesian period. They may have eventually fallen victim to ecological change brought on by forest clearance, or perhaps its eggs were devoured by the gluttonous kiore rat, introduced by early Maori settlers.

PREY NAÏVETÉ: DELUSION OR DRIVER?

Paul Martin's critics have pointed out that the prey naïveté argument does not pass a number of commonsense tests, nor does it accord very well with insights from recent animal behavior research. First, it is highly improbable that American megafaunal species could have been naïve in any overall sense, a point that Martin acknowledged. They had evolved within environments that included large carnivores, both mammalian and avian, with which they shared the landscape and to whose presence they were presumably well adapted. Just as is the case with extant species, these adaptations would have surely included the innate ability to quickly react to predators detected by olfactory and visual cues, or to respond to alarm signals emitted by other animals, including other species.

Second, if there is such a thing as a "generalized" prey response to threats, why would megafauna conditioned to native predators have failed to react automatically and appropriately to the new threat (i.e., humans) when it appeared? Like many other behaviors critical for survival, responsiveness to predator presence has a high if variable genetic component in prey species. Herding ungulates, with their highly developed social structures, for whom every sound or movement immediately draws their focused attention, would appear to be among the species least likely to behave naïvely by simply doing nothing when confronted by something unusual. Also, predator awareness isn't solely attributable to genetics or "instinct": among mammals, at least, learned responses can be transmitted intergenerationally, or integrated with other, more generalized reactions to danger.

Martin did not take a nuanced approach to characterizing differences in prey naïveté, but he did allow that inexperienced North American megafauna might have eventually figured out what humans were up to—had they managed to hang on long enough to do so. But in Martin's conception of faunal blitzkrieg, they never had a chance to learn. In his influential 1973 paper in *Science*, he posited that early paleohunters moved in a "bow-wave" or "front" of destruction across the Americas, where nothing was left behind and only the still naïve lay beyond (see figure 10.1). Along the rolling front there would be concentrations of "people" (i.e., social groups, not just hunters, averaging 40 persons per 100 sq km) engaged in hunting and processing the still naïve animals as they were encountered, moving at an average rate of 16 km per year. Within a few years, as Martin describes it, "the population of vulnerable large animals on the front would have been severely reduced or obliterated. As the fauna vanished, the front swept on, while any remaining human population would have been driven to seek new resources."

This latter group, which we can call the "left-behind" (averaging 4 persons per 100 sq km), would presumably have taken up more sustainable kinds of hunting and probably

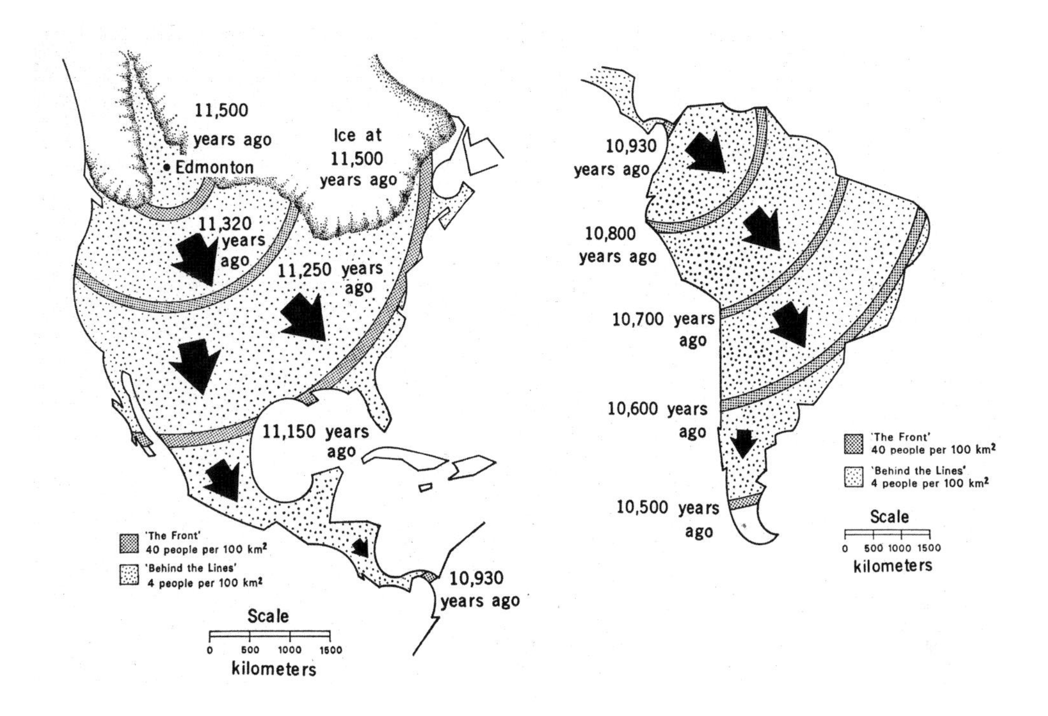

FIG. 10.1. **DISCOVERING AMERICA:** Martin's bow-wave of blitzkrieger paleohunters was supposed to have passed through the continental Americas like a veritable tsunami, traveling from the ice-free corridor (near present-day Edmonton) to southern Patagonia in approximately 1000 years. The dates appearing on the maps, from a famous 1973 paper by Martin are uncalibrated. With calibration, dates would be more in line with those given elsewhere in this book.

gathering. Further, both blitzkriegers and those left-behind would experience massive increases in numbers thanks to the amount of provender: Martin estimated a growth rate of 1.4 percent annually, attaining geographical "saturation" in 800 years.

Whether or not Martin was being metaphorical, the notion of the bow-wave was roundly criticized as being far too simplistic, if not completely unrealistic. It might have been possible for early human immigrants to move at a prodigious rate across open plains, but surely not through broken or rising ground, nor wetlands of any size, nor dense tropical forests, to name a few of the environments that they would have had to either negotiate or avoid as they spread through the Americas. Even allowing for the sake of argument that early paleohunters could have traveled rapidly through most of the physical settings they would have encountered, it is quite another thing to imagine that they could have reduced every megafaunal species along the way to such an extent that in the vast majority of cases there was no survivorship, no recovery *at all*.

From a biological standpoint, Martin's implicit requirement that persecuted species uniformly failed to recover because they were brought down to negligible numbers almost instantaneously as the front passed through the Americas, is, I think, one of the weakest links in the overkill argument. Martin and his supporters claimed that there were some examples of recovery, extant *Bison bison* being a particularly interesting case because it became the dominant large herbivore in Holocene North America. It is certainly plausible that bison might have benefited from the disappearance of mammoths and horses, as this would have reduced interspecies competition for the grassland habitats that all three favored (see plate 10.2). Also, near absence of competition would have allowed bison populations to expand dramatically, especially in the central and western parts of North America where no other herding herbivores in their size range existed after the Pleistocene–Holocene transition. But accounting for bison survival in this way simply makes the fact that other species didn't harder to understand. Why was there no horse survival, even if modest, in western North America, where mustangs do well today? Why didn't the western camel (*Camelops hesternus*), a species able to prosper in interglacial forests as well as arid regions, make it through to the Holocene in these same areas? The western camel was an artiodactyl (even-toed ungulate), the huge group that includes (in North America) bison, musk oxen, pronghorns, deer, moose, and so on—in short, all of the large herbivores that we are familiar with today on this continent. Although losses occurred within individual artiodactyl genera during the end-Pleistocene, the camelids were the only major lineage to completely disappear. Why? Bad adaptations, or just bad luck? Lack of answers to such questions is to be expected at one level, but it points up how little we know about the factors determining whether a given megafaunal species would be a winner or a loser in Near Time.

THE ISLAND SYNDROME

"Tameness" is a word with many shades of meaning. One set of meanings relates to the kind of behavior characteristic of domesticated animals that have been bred for thousands of years to respond reliably and tractably to humans, without aggressive or other unpleasant behaviors. In short, they've been genetically selected to express tameness, and we're the ones who've done the selecting.

Another set of meanings refers to wild animals whose behavior has been modified through intensive training, as in the case of performing seals or lions in old-time circuses. Here, the individual animal has learned to satisfy its trainer by balancing a ball on its nose or by jumping through a fiery hoop, in return for which it receives food or other inducements. Unlike the situation with the much more highly selected domestic dog, genetically enabled to predict and react to human behavior in a wide range of contexts, there's not much of an interspecies mind-meld going on with trained wild animals. It's more a question of

PLATE 10.2. **TARPAN OR WILD HORSE:** The ancestor of the living equids (horses, asses, and zebras) evolved in North America early in the Pliocene, about 5 million years ago. Its descendants ultimately spread to South America and Asia over land bridges. The modern domesticated horse, *Equus caballus*, is ultimately derived from wild Eurasian forebears through domestication experiments that started at least 6000 years ago. The existing wild horse of central Asia, *E. przewalski*, may represent this early stock, although this is disputed. So might the animals seen in European cave paintings, with their stocky builds, striped or stippled backs, and (possibly) erect manes. In any case, well before domestication began, modern humans were interacting with wild horses, for the latter's butchered bones are frequently encountered in Eurasian Paleolithic archeological sites. "Tarpan," a Turkic word for "wild horse," is sometimes applied to these animals. Yet, unlike the situation in North America, nothing ultimately happened: horses survived. In eastern Europe, tarpans survived into the nineteenth century, but these may have simply been feral animals (i.e., free-living domesticated horses).

FIG. 10.2. **TAMENESS KILLS:** The metallic pigeon is broadly distributed in eastern Indonesia, New Guinea, and elsewhere in the South Pacific. *Columba vitiensis griseogularis*, seen here, is a living subspecies native to the Philippines. It is closely related to the similarly colored Lord Howe pigeon (*C. v. godmanae*), which disappeared in the 1850s due to excessive persecution.

specific stimulus/narrow response, produced through constant repetition and reinforcement. Both the trainer and the trained learn how to behave in a stereotypic way to achieve a limited goal, without the trainer being eaten or the lion shot or otherwise abused in the process. Other factors may also be important, such as the ability of the trainer to take the alpha position in an implied dominance hierarchy. But unlike the case with domestic animals, programming tame behavior in a truly wild animal is due to learning: it is not facilitated by strong genetic predispositions developed through artificial selection.

There is, however, a third kind of relevant behavior in nondomestic animals. The hallmark is, once again, tameness, but in this case the behavior is not due to either individual learning or artificial selection. Instead, it is inherent, a by-product of a very different selective regime. In the literature it is often referred to as ecological or behavioral naïveté, and it is intimately connected with island life. Indeed, as David Quammen remarks in his book *The Song of the Dodo*, almost all of the historical examples of inherent tameness concern island endemics reacting to humans with astonishing passivity, in situations that would, in normal circumstances, be perceived as being fraught with imminent danger. Although this will take us out of prehistoric Near Time and into the modern era, this kind of naïveté is very relevant to the problem of understanding how innate behaviors may facilitate rather than prevent extinction.

A particularly ghastly instance of ecological naïveté concerns tiny Lord Howe Island (5.5 sq mi or 14 sq km) situated off eastern Australia, which is said to have lost more species and subspecies of birds (at least fifteen) during the last 500 years than Africa, Asia, and Europe combined. The island was first discovered in 1788 and visited occasionally thereafter—just enough to cause the demise of most of its bird life. The metallic pigeon, *Columba vitiensis*, a widespread species in the Pacific, was represented on Lord Howe by the now extinct subspecies *C. v. godmanae* (see figure 10.2). The bird's behavior when confronted by humans was both heartbreaking and incomprehensible, as ornithologist Jean-Christophe Balouet relates:

> [These] pigeons . . . were very plentiful and not afraid of man. Ships' crews had no difficulty in catching them on branches because they did not fly away. Once they were captured their legs were broken, and the cries of suffering from these unfortunate birds then attracted all others in the neighborhood, so they too were then taken.

With regard to mammals, a similar and only slightly less brutal story can be told of the Falkland Islands fox or warrah, a native canid first encountered by Europeans in the late seventeenth century (see figure 10.3). The warrah's scientific name, *Dusicyon australis*, means "foolish dog of the south," an all too appropriate label. The warrah was still extant when Darwin visited the islands during the voyage of the *Beagle* in the early 1830s, although its numbers were much depleted because its fur was valuable. Darwin marveled at the warrah's naïve, trusting behavior:

> These wolves are well known . . . [on] account of their tameness and curiosity; which the sailors, who ran into the water to avoid them, mistook for fierceness. To this day their manners remain the same. They have been observed to enter a tent, and actually pull some meat from beneath the head of a sleeping seaman. The Gauchos, also, have frequently killed them in the evening, by holding out a piece of meat in one hand, and in the other a knife ready to stick them. . . . Within a very few years . . . in all probability this fox will be classed with the dodo, as an animal which has perished from the face of the earth.

So what was going on with these island species that had rarely or never confronted humans before? Although most early modern-era reports are insufficiently detailed for any solid conclusions, it certainly looks as though island mammals and birds often failed to perform even the simplest avoidance maneuvers, such as just getting out of the way or hiding until the humans passed by. "Tameness" as a term does not begin to cover so disengaged a response to the unknown, let alone to an objective threat. Was this solely because these animals lacked any experience with predation or aggressive interactions over resources or territory? Or was it because their behavioral repertoires were so radically modified as a result of earlier, pervasive adaptations to island life that they had limited capacity to react to or learn from novel situations? Sadly, we will probably never know, although researchers are attempting to indirectly infer what might have happened, by looking for correlations in certain traits among quite different kinds of mammals.

For example, Marcelo Sánchez-Villagra of the University of Zurich and his colleagues have recently drawn attention to the "island syndrome," a term that refers to distinctive morphological and behavioral features found in various combinations in island species, such as shortened limbs, reduced sexual dimorphism, early maturation, and tameness. Using a wide array of indicators, the Zurich team thinks that such bundles of correlated traits could arise from the fixation (selection) of certain mutations affecting early embryonic development. A key example concerns the coordinated migration of certain cells (called neural crest) from one part of the embryo to another, where they are switched on and differentiate into specific tissues such as pigment cells, smooth muscle, and cranial skeletal structures. Mutations that affect differentiation may in turn affect many systems as development continues, in what is known as a cascade. Even if affected individuals survive, in the normal case they would potentially have reduced fitness compared to other individuals in the same species. However, for populations in the process of adapting to island life, the normal case may no longer apply if the fitness rules have been changed. Paleontological evidence shows that features that might have led to seriously reduced fitness on mainlands, such as tiny body size in elephants, may

FIG 10.3. **THE WARRAH** (*Dusicyon australis*), also called the Falkland Islands fox or wolf. It was neither of these, its closest relative being *Chrysocyon brachyurus*, the living South American maned wolf (also not a wolf). Charles Darwin, hearing about warrahs in 1834 during a visit to the islands, predicted that they would shortly go the way of the dodo. And so they did: 1876 was the last time a living animal was seen.

prove highly favorable in the harsh, resource-poor environments characteristic of many islands (see plate 10.3).

What gives this idea its special significance is the fact that domesticated species express similar sets of traits that likewise have no obvious connection with one another, but can be traced back to early developmental errors. This is the "domestication syndrome," which dif-

PLATE 10.3. **DWARF CYPRUS ELEPHANT, NO TALLER THAN A FALLOW DEER:** By comparison to living elephants, Mediterranean dwarf elephants like *Palaeoloxodon cypriotes* from Cyprus and *P. falconeri* from Malta (plate 4.6) may have lived "fast" lives. Modern African elephants become sexually mature around eleven to fourteen years of age, but according to some analyses Pleistocene dwarves developed much more rapidly, attaining sexual maturity at an estimated age of four in the case of the Maltese elephant. Average life span may have been a short twenty-five years, with pregnancies lasting just eleven or twelve months—half as long as in African elephants. Whether "fast" lives occurred in other insular species has not been established. Other studies have shown that downsizing could occur incredibly quickly if selection for small size was strong: dwarfism is highly adaptive where resources are reduced or uncertain, as is often the case in island environments. Interestingly, deer were just as successful as elephants and hippos in getting to Mediterranean islands. The one in this scene, however, is the Persian fallow deer (*Dama dama mesopotamica*). It was possibly introduced to Cyprus during Neolithic times, but did not manage to survive there.

fers from the island syndrome in that the trait packages expressed in domesticants are due to artificial selection for attributes like early maturation and reduced aggressiveness. It is this remarkable symmetry between the consequences of natural and artificial selection that gives the island syndrome hypothesis its explanatory strength, although whether ecological naïveté in island species and the tameness seen in domestic animals truly have a common source in developmental cascades needs further study. For the present, it provides a possible reason for why the warrah, Lord Howe pigeon, and dozens of other island species never "learned" to detect predation before it was too late: they may have been betrayed by the same selective mechanisms that enabled their successful adaptation to island life in the first place.

11

OTHER IDEAS: THE SEARCH CONTINUES

PLATE 11.1. **THE TOXODON, DARWIN'S "STRANGEST ANIMAL EVER DISCOVERED":** Rhino-sized *Toxodon platensis* was one of the last of the meridiungulates, a uniquely South American group that finally died out during the Near Time extinctions. Darwin, who collected the first fossils of this species during the *Beagle* expedition, once described it as "perhaps the strangest animal ever discovered" because it exhibited traits previously regarded as distinctive of species as diverse as elephants, rodents, and sea cows. Like *Macrauchenia* (plate 9.5), whose fossils he also discovered, Darwin couldn't place *Toxodon* anywhere on the mammalian family tree, and its peculiar grab bag of features made him think hard about his developing theory of evolution by means of natural selection. As he recorded in one of his notebooks dedicated to the "transmutation" of species, he had been "greatly struck . . . on character of S. American fossils—& species on Galapagos Archipelago.—These facts [are the] origin (especially latter) of all my views." There is no good evidence that *Toxodon* was hunted by early South American Indians.

At this point it might well be concluded that the classic contenders for explaining Near Time extinctions—climate change versus human persecution—have now reached a stage of intellectual impasse, not to say staleness, in which the few strong arguments that each can muster in its favor are essentially neutralized by the existence of numerous and all too obvious weaknesses. Yet the losses happened, and await an accounting. To do that, it is time to move the debate forward and to get away from viewing climate change and overkill as the exclusive choices for the cause of Near Time extinctions, as though all of them had to be due to the action of one or the other. To show that there is still boundless enthusiasm for resolving this greatest of all extinction puzzles, I will briefly mention three novel hypotheses that offer challenging ways of rethinking the problem of what happened, and why.

FOOD WEB DISRUPTION

Although size has its advantages, it also has its costs. True, basal metabolic rates (i.e., daily energy consumption) tend to be lower in large-bodied mammals, but because of their size they have to take in absolutely large amounts of food just to maintain themselves. This is crucial, especially if their diet consists of items requiring much digestive processing (e.g., grasses in ungulate diets). This brings up the interesting question of the relationship, if any, between grave interruption of a species' normal food supply and the likelihood that this will contribute to its extinction. To examine this notion properly we first need to explore the concept of the food web.

PLATE 11.2. **OR IS THIS THE STRANGEST MAMMAL EVER?** Or does it seem strange only because we know so little about it? The marsupial "tapir" (*Palorchestes azael*) was the terminal member of a distinctive Sahulian lineage related to *Diprotodon*, *Zygamaturus*, and other diprotodontids (plate 7.4). Its common name is based on the fact that the shape of its nasal aperture implies a fleshy proboscis (although probably not a functional trunk), perhaps something like that of a tapir. There is also a theory that it had an extensible tongue. That's already strange enough, but pony-sized *Palorchestes* also possessed elongated forearms and paws armed with large, compressed claws. Some paleontologists argue that this peculiar mix of adaptations implies that it used its powerful limbs to uproot tubers or bring down tree branches to get at leaves with that amazing tongue. Recently, paleontologists found a new and undistorted skull of this species, which revealed that its eye sockets were situated high on its head. This inspired the suggestion that it might have been rather hippolike in habit, spending much of its time in water (as tapirs do as well). This alternative idea is presented in plate 8.8.

The usual portrayal of trophic levels (or positions in a food chain) is the food pyramid, with predators on top feeding on herbivores that feed on plants and so on down to solar energy, the powerhouse that makes everything else possible. But this is a very simplified version of what really goes on. Ecologists prefer to emphasize the complexity of these energy-harvesting relationships with the more appropriate phrase "food web," which implies that such relationships actually constitute a tangle of multiple connections rather than a series of one-way streets.

Trying to infer how food web dynamics might have worked in extinct ecosystems is complicated for many reasons. For Quaternary studies, one such complication is the inherent limitations of the "deficient present," because so much biodiversity has been lost without replacement. Loss results in reduction of ecological complexity; and the greater the scale of loss, the more deficient the surviving portions of the ecosystem will appear. Thus in trying to reconstruct prehistoric food webs, we must be careful not to simply assume that they were more or less like those of the present. In regard to extinct species, we won't necessarily know who all the actors were or their roles, or what they would have consumed under different sets of conditions, or what behaviors might have helped or hindered their survival as conditions altered over time. Late Pleistocene environments seem ecologically unique to us today because they were dominated by megafaunal species that no longer exist (see plates 11.1 and 11.2). But in reality it is the deficient present that is unusual.

With these considerations in mind, Angel Segura, Richard Fariña, and Matías Arim of the Universidad de la República in Uruguay recently attempted to model how body size may have influenced food web collapse in Late Pleistocene South America. Body size is one of the few physiological parameters that can be estimated reasonably well for extinct species, permitting valid comparisons to modern ecological counterparts. The researchers divided up extinct species according to whether they could be considered carnivorous or noncarnivorous, and then cross-compared potential predator/prey sizes. To add reality to the model, they also calculated species vulnerability as a function of how many predators were in the correct size range to take prey of a given smaller body size. They then compared their results to known predator/prey size distributions in modern Africa.

Perhaps unsurprisingly, reconstructed Pleistocene food webs in South America were found not to be very different from modern ones in Africa, despite larger size ranges for both predator and prey. Only a few species in South America could have been entirely free from predation as adults, which mirrors conditions in modern Africa, where elephants and rhinos are the only mammals largely exempted from predation (save for human predation, of course). Nonetheless, according to the model in both places large predators would have tended to go after large prey, presumably because the cost of predation is so high that big payoffs are required.

However, as already noted, megafaunal species, whether these be lions and saber-tooth cats or elephants and ground sloths, have to eat—a lot. The system works well when diverse energy sources are readily available, but if they become sharply reduced for any rea-

PLATE 11.3. **THE LARGEST OF THE GIANT GROUND SLOTHS** *(Megatherium americanum)* may have weighed as much as 2000–4000 kg (4400–8800 lb). That makes it roughly 400 to 800 times the size of its extant and very distant relative, the three-toed sloth *Bradypus* (which weighs a mere 5 kg/11 lb). Comparably sized adult elephants must consume 100–300 kg (220–650 lb) of plant matter per day, spending most of their waking hours at the task. Unless giant sloths were different, significant interruptions in food supply or quality would have been devastating for them. Yet however much Pleistocene climate change affected their food supply, these huge sloths survived it all, as did mammoths, until anatomically modern peoples showed up. Standing carefully next to the sloth's hind foot is a marsh deer (*Blastocerus dichotomus*). It may seem tiny by comparison, but it is the largest South American deer species, with an average body size of 100 kg (220 lb).

son, so does the species' well-being. For carnivores and noncarnivores alike, the usual solution to scarcity is to move over to alternative if less valuable energy resources, such as less nutritious plants or smaller prey. The authors contend that such resource switching is natural and should not end in catastrophe if the fauna can cope with the overall stress brought on by scarcity. However, if additional stressors are introduced, things can go badly, and quickly. In South America 13,000 years ago, species at the upper end of their size distributions were already suffering from increasingly inadequate nutrition due to progressive aridification, forest fragmentation, and lowered plant productivity. Then humans appeared, with inevitable results. The authors conclude that "[a]ny shift in baseline conditions (e.g., resource availability) or the appearance of a novel predator able to hunt these organisms could lead [these] species into a highway to hell." This suggests that food web collapse was primary and human predation secondary, but with both disasters happening at the same time, the point is that the animals didn't have a chance. The arrival of human hunters was the final blow, tipping species over the edge. And we see the result: only a handful of the South American megafauna have managed to survive into the deficient present.

Climate-driven collapse of food webs provides a compelling reason for the apparent susceptibility of the South American megafauna to extinction, but were large species really so limited in their ability to cope with ecological change? That's another difficult but interesting question. As a thought experiment, paleontologist Richard Fariña made the radical suggestion that the giant ground sloth *Megatherium* (see plate 11.3), faced with an increasingly uncertain environment, could have taken on additional energy sources by becoming an opportunistic meat scavenger, on the model of the highly omnivorous brown bear (*Ursus arctos*) of North America. This species has made a success story out of being able to prosper on a diet of literally anything edible, including carrion. Although modern ecological research reveals that there is a certain amount of plasticity in a species' utilization of resources in a food web, for the vast majority of animals such plasticity as there is tends to work for only a limited period. When hard times become long-term reality, it's usually a question of either moving away or dying off. Whether *Megatherium* turned scavenger or not, in the long run it made no difference; like all the other very large mammals of the South American Pleistocene, it did not survive the Near Time extinction pulse, whether that was driven by environmental change, human depredation, or both factors.

Another radical suggestion which merits consideration is that collapse of species occupying one part of a large food web might produce knock-on effects that could contribute indirectly to extinction in other parts. That's the message accompanying Elin Whitney-Smith's second-order predation hypothesis, which posits that paleohunters in Late Pleistocene North America caused ecological disaster by eradicating native carnivores in order to reduce competition for favored prey. In doing so, they destroyed the natural balance then existing between predator and megaherbivore population sizes, instituting a boom-bust cycle in which unconstrained increase in prey numbers led to environmental exhaustion and, eventually, wholesale extinction.

Although some carefully constrained experimental studies indicate that boom-bust cycles of this sort can result in local extirpations, scaling up to continental dimensions and involving dozens or even hundreds of species in complex interactions is a difficult problem to model realistically. The ability of early American paleohunters to control predator population sizes with the technology available to them would surely have been a supremely chal-

PLATE 11.4. **SHORT-FACED BEAR AND SABERTOOTH CAT ARGUE OVER A BISON KILL.** The short-faced bear (*Arctodus simus*) was the largest carnivorous land mammal in Quaternary North America. It is estimated to have stood 2.5–3.5 m (8–11.5 ft) at the shoulders and weighed around 800 kg (2000 lb)—which is as large as an adult bison. Its unusual common name is based on the relative shortness of its muzzle compared to that of other bears. Extinct tremarctine bears also lived in South America, where they attained even greater size (plate 2.4). Ideas about their diets have fluctuated widely, from hypercarnivory to scavenging. In this painting, the standoff between the bear and the sabertooth may seem unlikely, but it is meant to reflect the view of some paleontologists that short-faced bears were basically scavengers, able to muscle in on the kills of other carnivores because of their great size.

lenging task except, again, over limited areas. Also, local resource exhaustion does not have the same consequences for mammals, which are highly mobile organisms, as it does for the kinds of microorganisms commonly used in ecological experiments. Migration might mean substantial range and population contraction, but that is not the same as an irrevocable death sentence for staying in place.

Finally, there just isn't any evidence that Late Pleistocene hunters in Siberia or North America began their occupation of new territories by intensively extirpating the kinds of large carnivores (large cats, wolves, bears, hyenas) with which they might have been in competition (see plate 11.4). Although food web disintegration must have occurred in the Late Pleistocene as a consequence of megaherbivore collapse, it remains as obscure as ever whether humans had anything to do with this, directly or indirectly. Unless strong evidence emerges that there were major predator-on-predator interactions in end-Pleistocene North America that preceded megaherbivore die-offs, second-order predation is an unlikely contender as an explanation for Near Time losses.

HYPERDISEASE

The disease hypothesis—or, to employ the more exclamatory name that Preston Marx and I proposed for it in 1997, hyperdisease—holds that emerging infectious diseases might, in certain circumstances, prove exceptionally lethal to populations and species and thereby cause or contribute to their extinction. At special risk would be species that lacked any kind of genetic or acquired resistance to the pathogens being introduced, as would often be the case at biological first contact.

The scenario that we developed was centered on migrating humans entering new lands for the first time and carrying whatever disease-inducing organisms that might have been associated with them (including vertebrate or invertebrate carriers, if any). The organisms of interest could be virtually anything that would have produced the hypothesized outcome—swift, nonrecoverable population collapse in escalating panzootics (i.e., diseases affecting many animals of one species, or animals of many species, over wide areas). In addition to mass mortality, the critical effect would have been overwhelming disruption of breeding behavior, interrupting the production of young. Although disruption might take several forms, the easiest to imagine would be excessive, rapid mortality, particularly among breeding-age individuals, but also among the very old and very young with compromised or underdeveloped immune systems.

The issue that the hyperdisease hypothesis was meant to address was how to explain the staggering abruptness of multispecies losses in North and South America as well as on the world's islands. To Paul Martin and his supporters, this could only mean that direct, intentional human involvement was responsible. To Preston Marx and me, it could mean human involvement of some sort, but in view of the rapidity with which the extinctions

FIG. 11.1

apparently took place, overhunting as the primary cause was just not plausible. There had to have been something else involved, perhaps something already known to have won countless wars of extermination in the annals of human history, without any direct involvement of arms or armies: our strongest ally, General Disease. In other words, hyperdisease is overkill, but without a conscious perpetrator (see figure 11.1).

In comparison to other anthropogenic theories, the disease hypothesis presented two theoretical advantages and one serious disadvantage. The first advantage was that it provided an alternative explanation for sudden losses after biological first contact, when humans and native animals met for the first time. It wasn't what people actually *did*, by way of hunting or other forms of overt persecution as in typical overkill arguments, but what they inadvertently *introduced*, diseases that took extremely rapid, fatal courses in previously unexposed populations.

In our scenario, the very limited number of instances in the archeological record in which there was actual evidence of Pleistocene megafaunal hunting could now be reinterpreted in a way that made more sense. In effect, such examples as there are do not represent a chance sampling of an otherwise immense slaughter, the existence of which has mostly been erased by time and the elements, but instead just what they appear to be: ordinary, occasional, limited efforts at big-game hunting by small bands of hunters and gatherers fulfilling their immediate needs. Under the terms of the hyperdisease hypothesis, it was no longer necessary to believe that tiny, unorganized bands could have somehow managed, all by themselves, to cause the deaths of millions of individual animals throughout the continental Americas in a geological instant. Marx and I were claiming that it was the effects of virulent emerging diseases, not blitzkrieg, that forced the losses.

Our hypothesis gained some credence from well examined instances of widespread population collapses due to disease in the modern era, such as the rinderpest epizootic in eastern Africa in the 1890s, which attacked most of the region's native even-toed ungulates and caused appalling mortality. Some species were very seriously affected, with one subspecies of hartebeest disappearing in the early twentieth century in possible correlation with the disease's outbreak. A more recent example of a disease-induced disaster was the die-off of more than 80 percent of the central Asiatic wild herd of saiga antelope (*Saiga tatarica*; see plate 8.7) in 2015–2016 from hemorrhagic septicemia, or blood poisoning due to bacterial infection. There are still other examples of almost unbelievable levels of mortality in wild animals within breathtakingly short intervals, all of which underlines the fact that there is really nothing in ordinary nature that can bring down the standing crop of a species as quickly as emerging infectious diseases.

The second advantage was that the disease hypothesis could account for the apparently nearly random survival of some megafauna. Although according to the hypothesis many individuals in many species would have been infected, not all would have been. If enough survivors were left, then despite having been squeezed through a very narrow bottleneck the species might recover, now with herd immunity to whatever disease brought down previous populations. The reason this is relevant is that it helps to explain an apparent anomaly in the blitzkrieg argument. In all continental areas where Near Time extinctions occurred, there was a sharp drop-off in species-level extinctions among mammals after the effects of biological first contact had run their course. In North and South America, for example, very few species-level extinctions are known to have occurred among terrestrial mammals, large or small, from the early part of the Holocene until the beginning of the modern era five hundred years ago. Populations surely occasionally disappeared, as in the case of some bison subspecies, but not entire species. In Australia the extinctions time line has been especially difficult to sort out, but the point remains that, after the losses in the Late Pleistocene, nothing much happened until the modern era, when the extinction rate shot up again. Clearly, something happened to drastically reduce the likelihood of outright extinction on the continents during most of the Holocene. Was that due to a change in human behavior, or to the reduction or elimination of the fundamental cause of loss?

The major weakness of the disease hypothesis concerns whether one or a few types of illnesses could seriously infect many different species at the same time. Although diseases are known that can and do infect a wide range of hosts, such as influenza, tuberculosis, leprosy, distemper, and many kinds of hemorrhagic fevers, the thought that any of them could have caused massive simultaneous infections in dozens of species, no matter how immunologically naïve, is admittedly hard to accept. Nor were we able to find empirical evidence of a prehistoric disease that fit the bill.

However, in one instance we were able to show that a specific disease had apparently caused during historical times the loss of two species of endemic rats on a small, isolated

FIG. 11.2. **KILLER PATHOGENS:** The two native rats of Christmas Island, the bulldog rat (*Rattus macleari*) and the Christmas Island rat (*Rattus nativitatis*, seen here), disappeared less than a decade after phosphate mining started on this previously uninhabited island. A medical officer, noting that the native rodents appeared to be rapidly dying out, wondered if their collapse might be due to murine trypanosomiasis, perhaps brought onto the island by introduced Eurasian black rats (*Rattus rattus*). It was an insightful guess: in 2008, scientists using ancient DNA techniques detected evidence of the hypothesized pathogen in several museum specimens of Christmas Island rats. Evidently, the island species had never previously experienced *Trypanosoma* infection and thus possessed no genetic immunity. This in turn led to unsustainable levels of mortality and, inevitably, extinction.

landmass in the Indian Ocean, Christmas Island (see fig. 11.2). The pathogen was a species of the protozoan *Trypanosoma* (a different species of which is responsible for human sleeping sickness, a potentially fatal disease). Trypanosomiasis was probably introduced by Eurasian black rats onto the island at the end of the 1890s; the endemic rats evidently had no immunity to infection and were gone within a decade or so.

Other scientists are using more advanced techniques to find genomic evidence of pathogens in ancient humans, and I look forward to the day when these approaches are applied to prehistoric populations of animals. For now, all that can really be said is that hyperdisease conditions may have prompted or contributed to some Near Time extinctions, but as a general explanation it remains unlikely.

FIREBALL

In the mid-2000s, nuclear scientist Richard Firestone and his colleagues presented an idea that frankly shattered all previous thinking about early Near Time extinctions. They argued that a 10-km-wide bolide, or cometary impactor, struck the Earth's atmosphere about 12,900 years ago. The authors conjectured that the bolide broke up over North America, sending a thermal pulse and associated shock wave that resulted in widespread fires and other environmental effects, which not only caused megafaunal extinctions and destroyed the Clovis culture, but also ushered in the Younger Dryas (see figure 11.3). There is no known impact crater, so the evidence for the event consists of various proxy indicators that, they claim, point to a calamitous visit from outer space. These indicators include distinctive microspherules, glass melts, nanodiamonds, elevated (thus non-Earth-like) levels of platinum-group metals, peculiar organic "black mats" of uncertain origin but frequently associated with Clovis-age contexts, and still other evidence that is, or at least could be, consistent with their hypothesis.

This was wild stuff, seemingly out of a movie script. Although in several later papers fireball enthusiasts dropped some of their earlier claims, they also provided additional evidence that they thought bolstered their argument that a cosmic collision had occurred. It hardly needs saying that the idea of a bolide forcing extinctions in the latest Pleistocene did not go over well with most Quaternary paleontologists and archeologists, who uniformly favored more down-to-Earth causes.

I remain agnostic about some of the ideas put forward by fireball supporters, but I respect their assessment of the indirect evidence for this cosmic visitor. I think they have adequately shown that there is good evidence for the existence of the bolide itself, although what exactly happened after it entered Earth's atmosphere remains elusive. In recent papers, fireball supporter James Kennett and his coauthors marshaled geological and chronological evidence from twenty-three widely separated sites that overlap the beginning of the Younger Dryas. They argued that there is sufficient site-to-site similarity to justify

FIG. 11.3

the conclusion that the fireball and the onset of the Younger Dryas were paired, the former causing the latter. Assuming that the latter's onset was also correlated with megafaunal extinctions in the continental Americas, here we have an argument that provides a significant alternative to all others that I have considered in this book. Or does it?

First, it is important to note that their most compelling evidence, in the form of microspherules and other tiny bits of cosmic debris, comes almost entirely from North America and western Eurasia. Only one site in South America—Mucuñuque, in northwestern Venezuela—has produced such evidence. The record as given is silent for both Africa and Australia, presumably because any signal for it would have been too attenuated to be detectable, although this is not stated. But that makes the geographical distribution of fireball markers all the more important. The Venezuelan site is in the Andes and classed as "lower quality" because the suspected Younger Dryas boundary layer produced no organic material that could be dated (although nearby and slightly older sediments did yield such material). Surprisingly, despite the existence of many well-published latest Pleistocene paleontological and archeological localities in the southern part of South America, none was included in their survey. Thus it is hard to evaluate whether the same kinds of catastrophic effects inferred for North America also took place in South America, which of course experienced dramatic losses at nearly the same time.

This same point concerning lack of data from relevant places applies to southern and eastern Africa, where dated paleontological sites of appropriate age are numerous. The omission of Australia is more understandable, as there are very few reliably dated Late Pleistocene localities. Still, for the fireball hypothesis to gain credibility, these other areas need to be investigated just as intensively as the Americas. If the event was truly of great magnitude, distant parts of the world should show evidence of it as well, whether or not it induced extinctions in such areas.

Second, in the original paper on the fireball hypothesis it was argued that the megafauna (and, presumably, Clovis peoples) died as a result of direct bombardment, the impactor having broken up into the equivalent of numerous cluster bombs when it entered the atmosphere. That particular kill mechanism was quickly abandoned by fireball supporters in favor of the bolide's thermal effects, which in their conception must have induced the greatest Heinrich cooling event in the entire Pleistocene, bringing on catastrophic ice-sheet melting, global temperature drops, and consequent biological extinctions. This more indirect cause of loss seems credible at first glance, but the known pattern of losses does not correlate well with the impactor's imputed effects. In terms of the marker of greatest interest here—sudden occurrence of numerous extinctions—most of the action was in North America and northern Eurasia. Nothing much happened in the Southern Hemisphere at the very end of the Pleistocene, with the conspicuous exception of South America. Why was that continent severely affected, when other landmasses—southern Africa, Madagascar, Australia, and the many islands that dot the Pacific—escaped unscathed? (Recent investigations do support the idea that, 12,900 years ago, immense amounts of glacial meltwater were suddenly released into the Atlantic and Arctic Oceans, possibly shutting down deep-water circulation, but for completely different reasons, not because a bolide vaporized part of the North American ice sheet.)

Third, there is the question of who was affected, and how. The cold snap at the start of the Younger Dryas was clearly of great magnitude, since it essentially threw Earth back into the Last Glacial Maximum. Pollen evidence shows that this Mother of All Chills had a rapid, catastrophic effect on vegetation at higher latitudes, turning recovering landscapes into bleak tundra once again. Under these circumstances both herbivores and carnivores would have been acutely stressed, as their food webs collapsed around them. So, surely, given environmental change of this magnitude, can we not point to the onset of the Younger Dryas as the driver of at least some of the megafaunal extinctions?

Yes, the Younger Dryas certainly sounds like an environmental disaster of the correct magnitude to force great losses, which is why it has long been a favored cause of extinction in climate change arguments. However, ancient DNA specialists Alan Cooper, Jessica Metcalf, and their collaborators have recently pointed out that the continental New World extinctions are better correlated with the beginning of *warming* cycles. Genomic data for several megafaunal species indicates that population sizes grew smaller going into warm periods, then recovered during subsequent cooler intervals. Why things should have happened in this counterintuitive way is still obscure, even if (as this team contends) human impacts came on at the same time: the "lack of evidence for larger-scale ecological regime shifts during earlier periods [i.e., before radiocarbon time] when interstadial events were common, but modern humans were not, supports a synergistic role for humans in exacerbating the impacts of climate change and extinction in the terminal Pleistocene events." Africa, always the outlier, should not be forgotten when framing such generalizations. Modern humans existed on that continent all the way through the Late Pleistocene, but seemingly never assumed a "synergistic" role in provoking extinction, no matter how much climate change was going on.

I chose these alternative explanations for Near Time extinctions outlined in this chapter because they each have something to offer that the conventional arguments do not. They present challenging ideas that overkill and climate change theories have overlooked or failed to incorporate. Even more interestingly, they tie together facts from both sides of the original debate, giving them new explanatory contexts and requiring that we think again about what we think we know. That none satisfy all of the objections that can be mounted against them may be taken as a measure of how little we still understand, or alternatively how much more there is to discover, to test, to integrate.

I hope I have made it clear that the debate over the cause of the Near Time extinctions is not solely, or even mainly, about what may or may not have happened long ago and far away to this or that group of large animals. It is also about what we can actually hope to learn about such events in the distant past, for which the clues are scanty and the evidence, such as it is, appears pathetically inadequate for solving the problem at hand. About how apparently relevant data is collected and evaluated, and the role of assumptions and biases in framing and testing arguments. About imagining the effects of factors that have no referent within modern human experience, and the importance of insight in science. This is what continues to make the journey toward understanding worthwhile, whether or not the destination is achieved any time soon.

12

EXTINCTION MATTERS

PLATE 12.1. **HAAST'S EAGLE FIGHTS BACK**: *Harpagornis moorei* may have been the largest raptor in the world before it disappeared about 1400 CE. The eagle lived on the South Island of New Zealand and appears to have been an active predator, preying on moa and other birds. In this role, it would have had no competition, since mammalian carnivores were absent. Individual birds may have had wingspans as much as 2.3 m (7.5 ft), with body sizes of 10–15 kg (22–33 lb). (The largest living raptors are only two-thirds this size.) Unlike many island birds, Haast's eagle did not undergo wing reduction, despite its large size. Perhaps, as has been suggested, it used its body mass to stun prey by dive-bombing them. Although there is no direct evidence of the eagle attacking humans, there is a Maori legend of a very large bird (known as *pouakai*) doing just that.

WHO IS RIGHT?

It will probably come as no surprise when I say that scientists are quite mixed in their views regarding prevailing explanations for Near Time extinctions. In my very informal polling of colleagues who were willing to express an opinion, I found that many ecologists and most conservation biologists have no doubt at all that Paul Martin was correct in perceiving humanity's cloven hoofprints stamped all over the record of prehistoric extinctions. That does not mean that they necessarily accept overhunting as the exclusive kill mechanism. Indeed, given the lack of compelling evidence, few do, preferring instead to lay the blame on contributory activities, such as overexploitation, environmental damage, and the introduction of exotic species. In short, they know what they see; if the leading threat to life at present is us, then surely it was no different in the past 50,000 years.

By contrast, I find that many paleontologists and archeologists are reluctant to implicate human practices, arguing that factors other than overhunting must have been at work during Near Time, and in fact were likely the predominant causes of many or even most of these extinctions prior to 1500 CE. Archeologists specializing in New World prehistory vary, but many tend to strongly disagree with the notion that ancient big-game hunters could have overhunted on such a scale that they caused or initiated dozens and dozens of losses. Contributed, perhaps, but caused them singlehandedly? No. In Australia as well, judging from the recent literature most researchers do not support overkill, preferring to view climate and ecological change as the real culprit for the wave of extinctions on that continent 40,000 years ago. Of course, human agency can always be brought in to deliver a fatal blow to last survivors, but that becomes a tired trope after a while, especially when unsupported by any pertinent facts.

But perceptions are important, and it should not be lost sight of that the megafaunal extinction debate has a real bearing on how we think about current extinctions (see box, page 179). In effect, Near Time losses have become the opening act for the chief ecological morality play of our times, in which humanity is offered the possibility of redemption—should we manage to halt the Sixth Mass Extinction—for our past sins against nature. Despite the limitations of the overkill hypothesis, the view that humans were uniquely responsible for most Near Time losses is part of our zeitgeist, for reasons that are easy to understand. As the evidence for anthropogenic factors in our current spate of extinctions has grown, setting off alarm bells concerning still more losses in the near future, popular publications and the media frequently assume that humans must have been complicit in prehistoric losses as well. Whether this is a fair assessment or not, it needs to be confronted, as this book has attempted to do.

NEAR TIME EXTINCTIONS AND THE SIXTH MASS EXTINCTION

The "Sixth Mass Extinction" refers to the massive bout of species endangerment and loss the planet is currently experiencing, as a result of climate change and other detrimental transformations for which humans are overwhelmingly responsible. Some commentators think the roots of the Sixth run much deeper than a few centuries, and would incorporate all Near Time extinctions as the lead-in (or warm-up, as it were) to this looming disaster. From this standpoint, Near Time extinctions have thus never ended. Yet the Sixth Mass Extinction, if it takes place at levels currently predicted, will not simply replicate the extinctions of Near Time at a higher order of magnitude. It will have its own attributes, just as did the preceding Big Five. Human overhunting, in the specific sense discussed in this book, is not going to be a primary cause of large-scale losses, although some species will surely be lost to the bush-meat trade and poaching. Instead, it will be all that the phrase "human impacts" portends in the form of indirect but highly destructive, multicomponent threats, including overexploitation, loss of habitat, pollution, and induced climate change. There is nothing to be done about extinctions that have already occurred, except, of course, to learn as much as possible regarding why they happened so that this knowledge can serve as a framework for future action. That is, in fact, our only alternative.

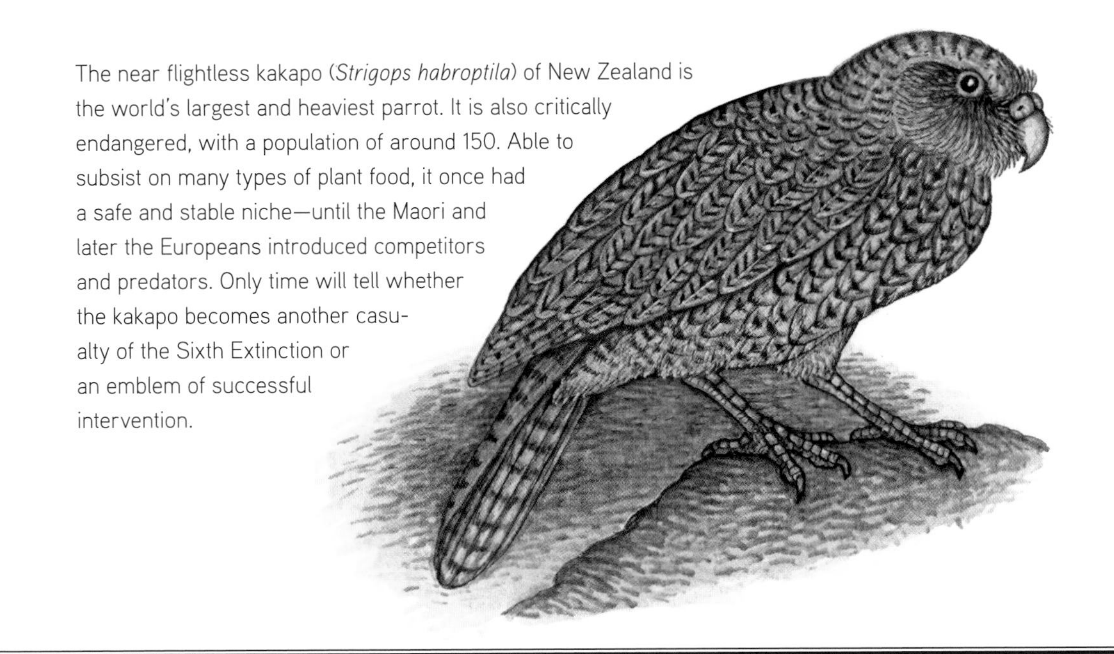

The near flightless kakapo (*Strigops habroptila*) of New Zealand is the world's largest and heaviest parrot. It is also critically endangered, with a population of around 150. Able to subsist on many types of plant food, it once had a safe and stable niche—until the Maori and later the Europeans introduced competitors and predators. Only time will tell whether the kakapo becomes another casualty of the Sixth Extinction or an emblem of successful intervention.

ONE CAUSE DOES NOT FIT ALL

In the preface, I promised that I would end this book with a sense of where the debate over the cause of Near Time extinctions stands at the moment—from my perspective, of course. It's now time to come to some conclusions about the leading theories we have examined, as well as to suggest some future avenues for rumination and research.

The part of the megafaunal extinction puzzle easiest to solve may be why large-bodied mammal species were especially unlucky in the specific circumstances of Near Time. As we have seen, living very large means low reproduction rates overall, with relatively few offspring per female reaching reproductive maturity (but see box, page 181). Any major disruption of population structure, such as might occur with greatly heightened infant and juvenile mortality, would have a proportionately greater effect on larger species than small ones, which tend to produce many more offspring per birthing interval. This, however, is an outcome; it does not tell us what the root cause of loss may have been, nor why it was particularly bad to be big during Near Time.

Martin's most important insight concerning these extinctions, that concentrated faunal disappearances did not occur until after the first arrival of people in affected areas, has not been definitively refuted in places where it has been meaningfully tested. However, except for a few island contexts (see plate 12.2), neither has it been overwhelmingly supported. For some extinction events, like the ones in North and South America, Martin's overkill hypothesis seems increasingly shaky, as earlier and earlier dates keep turning up for the start of human migration into the New World. Earlier presence implies longer periods of coexistence between people and prey, and in the logic of blitzkrieg that makes extinction less, not more, likely to occur.

Not many archeologists think it probable that the ancestors of native Americans could have entered the Western Hemisphere much before the start of the Last Glacial Maximum 27,500 years ago. If anyone was around that early in the Americas, their activities did not cause appreciable extinction. Thus proof that hominins of some sort were in North America as early as the Last Interglaciation 130,000 years ago would be truly revolutionary and would require rethinking the whole human enterprise in the New World, let alone their role in forcing extinctions. The only cultural evidence for so early a date comes from the Cerutti mastodon site in southern California, and it is slim indeed: some fractured bones and slightly modified rocks that may have been brought in from elsewhere by human (?) transport. Fast-forward 120,000 years to the very end of the Pleistocene, and we find that *Mammut americanum* was still around, even if its days were numbered. So lengthy a period for humans to interact with their prey does not suggest the operation of blitzkrieg within Martin's meaning of the word. If anything, it implies *nichtskrieg* ("nothing war")—that is, no more than ordinary levels of hunting, as practiced by hunters and gatherers throughout human history, a form of resource exploitation that does not result in the elimination of prey species.

REPTILE EXTINCTIONS IN NEAR TIME

Reptiles and amphibians have tended to receive much less attention than birds and mammal in studies of Near Time extinctions. For amphibians the scale of prehistoric loss is still very uncertain, but for reptiles the record has been greatly improved in the last several years, thanks to researchers making a concerted effort to identify when, where, and possibly how they came to grief. Zoologist Alex Slavenko of Tel Aviv University and his coworkers have shown that reptile diversity in Near Time was impacted in much the same way as that of birds and mammals: (1) losses on islands greatly predominated; (2) body size was important, but not equally so in all reptile groups; and (3) especially in the case of very recent losses, human-induced environmental perturbation was an overriding factor. Extinctions struck large lizards, turtles, and crocodiles hard, but large snakes less so. Large reptiles may have been prone to extinction because of their comparatively slow life histories, which lessened their ability to recover from human depredation. However, size wasn't everything. Large lizards, for example, typically lay larger, rather than smaller, clutches of eggs, so low fecundity was probably not the extinction trigger it seems to have been for large mammals. Among the lost megareptiles depicted in this book are the lizard *Varanus priscus* (plate 2.2), the snake *Wonambi naracoortensis* (plate 4.4), and the tortoise *Meiolania* sp. (plate 8.9), all of which lived in Sahul.

Lonesome George (?1910–24 June 2012), the last Pinta Island tortoise (*Chelonoidis abingdonii*), now on display at the Charles Darwin Research Station on Santa Cruz Island (Galapagos). Body sizes of as much as 400 kg (880 lb) and shell lengths of 135 cm (4.5 ft) have been reported for other Galapagos tortoises. Here a sense of scale is provided by the hitchhiker, an adult eastern box turtle (*Terrapene carolina carolina*) with a shell length of 13 cm (5 in).

If humans were definitely present that early—and only additional, unimpeachable instances of their occurrence will tell—then blitzkrieg, the only strong candidate that has ever been proposed as a direct anthropogenic kill mechanism, would be decisively falsified as the fundamental cause of New World extinctions.

NO COMMON CAUSE?

Common-cause explanations like the overkill hypothesis tend to be attractive because they are both reductive and predictive: given a specific set of facts, what happens next is inevitable. The facts may be noisy and even contradictory, but as long as the ones regarded as essential can be positioned in the proper order, the preferred explanation holds—or at any rate can be said to do a better job of imposing meaning than its competitors do. But one needs to be wary. Martin pushed overkill very hard as a common-cause explanation for Near Time extinctions, yet it has become increasingly clear that not all such losses followed precisely the same script. Martin thought that the continental New World at the end of the Pleistocene could serve as a kind of template for what repeatedly happened elsewhere when prehistoric peoples first arrived somewhere new. But does it? People of one sort or another were present in Eurasia for several hundred thousand years without causing mass losses. Furthermore, the losses that did occur were not necessarily fast. Isolated groups of woolly mammoths and giant Irish deer managed to survive into the mid-Holocene, albeit with vastly reduced ranges and population sizes, before finally dying out. A few other megafauna, including the extant musk ox, also managed to hang on in places where they no longer naturally occur, such as northern Siberia, until very late in the Holocene before succumbing. In this regard, the late Quaternary extinctions in northern as well as southern Eurasia look to some degree like those that occurred in nonblitzkrieg Africa, where extinctions were either scattered or prolonged, depending on one's perspective.

Australia might also be off script, since it now seems that people might have dutifully waited 20,000 years or more after their arrival in Sahul before initiating extinctions—if they did. If, by contrast, intensifying aridification resulted in at least sixty species-level losses in Australia spread out over as much as 20,000 years, as Steve Wroe of Australia's University of New England and his colleagues maintain, then all that can be said is that this was another sui generis event that permits no wider comparisons.

The truth is that both overkill and climate change suffer from plausibility issues concerning their effectiveness as single drivers of extinction. More than anything, it is the magnitude of the losses they supposedly forced during Near Time that strains our credulity. It is possible to imagine how, over a short period of time, small groups of hunters could deleteriously affect a species, or even several species, badly enough to lower standing population sizes. The same outcome could also occur in the case of severe climate change, if the youngest and oldest members of a species were especially hard hit by reductions in, say, food supply. But as species become rarer or less naïve, or as selection increasingly favors the hardier genes of those that remain, these drivers should become progressively less effective. Yet the megafauna, or at least most of them, died. The conundrum persists.

THE WAY FORWARD

Future progress in understanding Near Time extinctions will require the collection and manipulation of vast amounts of data, from many more sources than archeologists and paleontologists have customarily considered. Although looking for causal relationships within big data is a promising research direction, I emphasize that good ideas do not arise solely from large flocks of points on a diagram or simulations of what might have happened. As always, the prime objective scientifically is developing and testing ideas, so that, one day, an elusive insight can become a hard explanation. For now, we have to be content with outlining that which we think we know versus what we don't (but would very much like to):

1. To understand the similarities and differences between continental and island extinctions, we need a much better handle on the spread of modern humans through both kinds of geographical settings (e.g., plate 12.2). Dates of initial arrival are still inadequately known, or are no better than minimal estimates, for many of the places to which modern humans migrated. Although modeling suggests that some prehistoric extinctions could have been effected by small groups of humans acting very quickly, there is clearly no rule involved. And things do change with new and better evidence. When Martin was writing in the 1960s, Madagascar was regarded as a definite case of extremely rapid overkill. With new dates that double the length of human tenure on the island, we can no longer be quite so sure.
2. Population dynamics for the vast majority of species that became extinct in Near Time are unknown. For example, we know little or nothing about how they reacted to earlier climatic or environmental changes that may have predisposed them to extinction. The very few species for which we have genome-level information at present give mixed signals about whether they were stable, increasing, or in decline before human arrival. Without reliable information on species' vital statistics, any assumptions regarding their susceptibilities to either people or climate are simply guesswork. Yet despite its limitations, paleogenomics is the one science that can potentially provide such evidence, and thereby turn guesses into testable hypotheses.
3. Arguments that significantly depend on the formula "absence of evidence is evidence of absence" may sound definitive but they are often empty of real meaning. The truth is that Paul Martin would have been delighted had the remains of American megafauna been found in great abundance in sites dated to 12,000 to 13,000 years ago. He tried to transform the fact that kill sites are exceedingly rare from a weakness of the overkill

PLATE 12.2. **FIJIAN CROCODILE AMBUSHES A MEGAPODE:** Mammalian carnivores have rarely been successful in naturally establishing themselves on islands. Their absence has therefore given much scope to nonmammals, especially crocodiles and raptorial birds, to fill the available niche space. A now extinct crocodile group, the mekosuchians, lived in mainland Australia (plate 8.8) as well as on South Pacific islands like Fiji, Vanuatu, and New Caledonia, where they acted as top predators. Although poorly known paleontologically, some native crocodiles were still extant on Fiji when people arrived 3000 to 4000 years ago—but not for long. The species seen here, *Volia athollandersoni*, was part of a fauna that included a giant frog, a turtle, an iguana, and several native birds, including the deep-billed megapode (*Megavitiornis altirostris*), which is depicted here trying to make a run for it. All became extinct after human arrival, although how fast this happened is uncertain.

FIG 12.1. **AVIAN ARMAGGEDON:** As many as 2000 species and distinct populations of birds may have been lost during the last 800 years in the Pacific region. Many more are highly endangered, including the ones seen here, from the island of Kauai (counterclockwise from top: the ou, the akialoa, and the o'o). These birds, representing groups unique to Hawaii, have been hit hard by centuries of human persecution, environmental transformation, introduced species, and emerging diseases. During Near Time, a terrible fate has been visited on faunas everywhere, nowhere more obviously than on islands. The Earth too is ultimately an island; that should be remembered when we contemplate the future of what remains, and how it should be cared for.

hypothesis into a strength by asserting that near absence must mean that the extinctions happened astoundingly quickly. I sympathize with his conclusion but not its premise. It's not just the lack of kill sites, it's the lack of any indication of interactions with most of the disappeared *at all* that is the paramount absence in need of explanation. Pointing to the tiny number of sites in North America where limited hunting of species that are now extinct took place at the end of the Pleistocene does not alter that overriding reality. Absence is absence, full stop.

4. Climate change as a cause of extinction has its own issues to confront. There are arguably no more than two or three times during the past 50,000 years when major species losses

can be tightly matched up with truly major, rapid changes in climate, and even in these cases the kill mechanisms remain obscure. It is surely astounding that the largest climate event of all in the Late Pleistocene, the Last Glacial Maximum, seems to have caused negligible losses in the Northern Hemisphere, the part of the planet where its overall effects ought to have been the strongest and most detrimental. In the roller-coaster series of oscillations that occurred at the end of the Pleistocene, climate transitioned from one of gradual warming, to extreme cold, and then back to very warm over a period of less than two millennia (i.e., Allerød through Younger Dryas to Holocene; see figure 3.6). Although the effects of such changes were doubtless dramatic, their impacts outside of the Western Hemisphere and northern Eurasia were seemingly inconsequential, at least as far as large-scale extinction is concerned. To me this underlines the universal problem with climate-driven explanations: to be credible as a cause of previous extinction in a given large geographical context, the change had to be of great magnitude. This in itself should have set in motion many knock-on effects, provoking collapses in many places geographically remote from one another. This didn't happen at scale during the transitions just mentioned. However, it should be noted that traditional paleontological methods are acutely limited in this regard, for they can only detect outright extinctions, not near collapses from which species later rebounded, to survive for another day or perhaps to crash again. Perhaps someday paleogenomics will be able to say whether major population reductions, just short of full extinction, took place in places like Africa or South Asia 12,000 to 13,000 years ago. This would be an absolute game changer, forcing a thorough reappraisal of how we think about the causation of Near Time losses and the resilience of species.

But all of this is as it may be. Right now we have to play the cards we've been dealt, and none of the hands we've looked at so far are clear winners. So, if neither of the classic explanations for Late Pleistocene extinctions are tenable as general explanations, as I have tried to underline throughout this book, can they do the job in combination? A case can surely be made for assuming that cofactors of various kinds must have played a role in most Near Time extinctions, just as they do in modern losses (see figure 12.1). However, this isn't of much help if the cofactors cannot be separately distinguished and evaluated. Thus, in some contexts in which human depredations are thought to have occurred, environmental change as a cofactor also makes sense—but only in the form of catastrophic changes measured in years or decades, not as a slow flux brought on over the course of centuries or millennia. Emerging infectious diseases inadvertently introduced in the course of human migrations may have stricken a number of species in certain circumstances, but it has yet to be determined what those would have been. Especially in the case of islands, it may have been adaptations to insular life itself that predisposed species to extinction, where "tameness" ultimately became a death sentence. There are doubtless still other possibilities, yet to be revealed, which is why the study of Near Time extinctions continues to be such a field of intellectual ferment.

EPILOGUE: CAN THE MEGAFAUNA LIVE AGAIN?

Readers may remember the excitement at the turn of the century when it was proposed to "clone the mammoth." Fueled by a series of Discovery Channel specials, the idea was that all you needed to do was to find and sample a well-preserved woolly mammoth in Siberian permafrost, then get to work in the genetics lab. Alas, the hard lesson was that living cells are required for cloning, because they have intact DNA. Dead ones, no matter how well preserved, contain only shattered, degraded remnants of genomic material that cannot be used for this purpose. Even with excellent fresh material, cloning is still difficult. Reproductive biologist Ian Wilmot and his team at the Roslyn Institute in Scotland conducted more than four hundred experiments with enucleated eggs into which nuclei from adult cells had been transferred before they had their first success in 1996 with Dolly the Sheep. True, advances made by the cloning industry in the past two decades have made it easier to ensure that the genes of favorite pet dogs and high-producing cows will live on, but even so, living tissues are needed. Quixotic efforts to find mammoth mummies with viable cells that have somehow been kept in suspended animation for ten millennia are still being made in Siberia. So far, no luck.

This brings us to synthetic biology and genetic engineering, the most recent technology to offer the hope of bringing back the past. Imagine that you have two species, one extant, from which you can retrieve almost all of its genetic coding, and the other an extinct close relative. Although the latter's DNA will be degraded, with a lot of work it is now possible to acquire large parts of its total genome using ancient DNA techniques. What makes the next step feasible is the close relationship of the two species. They will be genetically very similar, which means that it should be possible to compare the two genomes, gene for gene, and determine where the few differences actually lie. Matches won't necessarily be perfect because of holes in the extinct species' information due to degradation, but by using the known genome of the extant species as a kind of scaffolding it should be good enough to embark on the third step. This involves making changes to individual genes in the genetic material of the extant species so that it matches that of the extinct species. To do this, scientists use gene editing tools (CRISPR/Cas9 is the current example) to strip out specific genetic sequences from the DNA of the extant species, which they replace using other tools with sequences retrieved from the extinct species. If the reworked genome can be introduced into gametes (sperms or eggs), then it should be possible to produce viable hybrid embryos that will, if everything goes well, develop into fully functional beings. The hybrids will express features of both parental organisms, but by careful selection over the course of several generations it should also be possible to get the desired combination of characteristics in a pure breeding line. That's the theory.

This very project is being undertaken right now in an effort to restore the woolly mammoth, and the chances of success are better than one might think. First, there is good information on the mammoth's genome, thanks to years of dedicated work by ancient DNA specialists. Second, there is an extant close relative, the Asian elephant. The woolly mammoth and the Asian elephant differ genetically from each other by only a few percentage points, much as chimpanzees differ from humans. Of course, the differences that do exist are critical, and detecting and managing them properly will distinguish success from failure.

Those who back this project hope that there will eventually be a pure breeding line of... what shall we call it? A "mammophant"? An "elemoth"? Presumably, the engineered version will closely approach the extinct form for the kinds of characteristics that we know from mammoth fossils, such as the big head, curvy tusks, and long furry coat, but it may differ in unknown ways for physiological processes and other characteristics that don't fossilize. This may not matter, at least to those who most want to see it: a mammophant that possesses, say, 80 percent of the mammoth genome may be phenotypically enough like our vision of the latter so that only the absurdly fastidious would insist it was not a real, live woolly.

But fantasies aside, for the foreseeable future there will likely be a number of limitations to what can be accomplished with genetic engineering. One of the considerations is the need for a closely related surrogate to birth the hybrid. There are several reasons for this requirement, an important one being that mammalian offspring are born with underdeveloped immune systems and depend on receiving antibodies in mother's milk. The antibodies enter the infant's circulatory system through the gut epithelium, and in addition to their protective role they also stimulate the infant's immune system. In the absence of such protection, newborn animals may fall prey to infectious diseases. Placing a clone or hybrid of a rare species into a surrogate mother from another, common species may ensure that this transfer takes place.

As long as surrogates, as opposed to pregnancy machines outfitted with synthetic placentas, are part of the operation there will still be a need for matchups of other sorts. That's why bringing a mammoth fetus to term in an Asian elephant surrogate makes sense: the body size of the mammoth baby will probably be similar to that of a normal Asian elephant baby. It's anybody's guess how close the natural behavior of mammoths was to that of living Asian elephants, but the hybrids will necessarily be socialized as Asian elephants.

By contrast, none of this would succeed with extinct ground sloths. I work on this group, and part of me would love to see a de-extincted *Megatherium*, for example (see plate 11.3). But adult *Megatherium* weighed a ton (actually, several tons), whereas the largest species of living tree sloths comes in at less than 5 kg (11 lb). Were it even possible to create a hybrid *Megatherium*, a full-term fetus would be far too large for surrogacy to be feasible.

Although there are as yet no newborn mammophants for us to adore—surely the ultimate in designer pet-babies—geneticist George Church of Harvard University and his colleagues are doing their best to make this a reality in the near future. Other scientists are aiming to create a simulacrum of the passenger pigeon, using genetic sequences collected from museum specimens that will be imported into the germ line of a close relative, the rock pigeon. Still others are bringing back the American chestnut tree by genetically engineering its genome so that new transgenic plants will be resistant to the fungal blight that killed off literally billions of mature chestnut trees across the country during the twentieth century. This kind of "facilitated adaptation," or the introduction of new genetic material into a species to increase genetic variation and thereby its ability to respond to challenges in its environment, might help to ensure the survival of species perched on the brink of extinction. As successes are scored and techniques improved, we can expect to see many interventions of this sort.

Another route to bringing back the past, sort of, is through "rewilding," one of Paul Martin's favorite ideas. Although the Pleistocene megafauna of North America are no longer with us, couldn't we have some pretty good stand-ins by importing their close relatives or ecological semiequivalents? Lions and cheetahs could make up for vanished sabertooths and giant jaguars, while Asian and African elephants might stand in for mammoths. In this way we might be able to restore ecological functions and relationships that were sundered by the megafaunal extinctions, and derive other benefits besides. In Australia, enthusiasts have advocated bringing in rhinos to fill the void left by the disappearance of huge herbivores like *Diprotodon*. In Europe, more modest proposals to introduce lynx and beavers, locally extinct for hundreds of years, compete for attention with those advocating the return of wolves and possibly other large carnivores that have lost half or more of their historical ranges thanks to human encroachment.

Researcher Sergey Zimov proposes to actually re-create the ecosystem of the mammoth steppe in the Kolyma basin of the Russian Far East, populating it with bison, horses, and other large herbivores whose activities might, over time, transform boggy taiga into productive grassland. Perhaps Church's hybrids will carry enough of the woolly's adaptations for extreme cold to survive in Zimov's Pleistocene Park, and maybe they could help with the intended transformation of parts of northern Siberia into rich grasslands. At one level I applaud his well-intentioned efforts, but how many such animals will be needed to make any difference to the plant cover, and who will take care of them over the long run? Are they to be free-ranging, or kept in enclosures? Would this really be rewilding, or just rezooing? Like overkill, it all makes a certain amount of sense up to a point, but then it falls off the hard edge of reality.

De-extinction is fascinating. There are numerous ethical issues to work through before unrestrained endorsement is appropriate. If the resurrected are to have real existences, should there not be equally intense efforts to provide them with spaces where they can live out their lives, not just in Arctic Siberia but elsewhere? If they are introduced into nature, what will this mean for other species living in the same environments in terms of competition, disease transmission, or other problems? These are all important questions with few answers at present. Whatever the outcome, it is to be hoped that any brave new creations can be integrated into existing ecosystems without destroying them—something we have been unable to do with ourselves, for at least the last 50,000 years.

APPENDIX:
DATING NEAR TIME

Radiocarbon and, to a lesser extent, optically stimulated luminescence dating are the major methods of acquiring information about the age of things in the Quaternary. There are many other methods that have useful applications for specific dating problems, but since they do not (yet) play a significant role in the story of Near Term extinctions, they will be omitted from review here.

Unless otherwise indicated, most dates in this book are based on the results of radiocarbon analyses, calibrated to calendar years. Most are also rounded off for simplicity, without error ranges. For example, credible last appearance (terminal) dates for North American species lost during the Pleistocene–Holocene transition mostly fall within a range of 10,500 to 11,000 radiocarbon years BP. Using a specific set of assumptions, this interval calibrates to a somewhat broader span in calendar years, 12,000 to 13,000 calendar years ago.

RADIOCARBON DATING

Carbon-14 is a radioactive isotope of carbon that can be used (within limits) to establish the probable age of anything containing measurable quantities of it. Until reliable calibration methods were developed, beginning in the 1980s and 1990s, it was generally assumed—wrongly—that there was a reasonable correspondence throughout between radiocarbon years and calendar years. This is generally correct for most of the Holocene; however, in the Late Pleistocene and Pleistocene–Holocene transition things get complicated, with unexplained anomalies that may or may not have something to do with the amount of carbon-14 being formed in the upper atmosphere during this interval. For this reason, I prefer to cite dates based on radiocarbon results in terms of their equivalents in calibrated calendar years. It does not solve the problem, but it makes it less confusing if there is just one kind of chronology in the text.

OPTICALLY STIMULATED LUMINESCENCE

This dating technique (abbreviated in the text as OSL) can be used to date the last time individual grains of the minerals quartz and feldspar were exposed to sunlight—that is, before they were last buried in sediment. The method, which does not depend on the decay of a radioactive isotope, is very useful because it dates the actual sediments in which fossils were deposited. (Quartz and feldspar are the two most abundant minerals on the Earth's surface.) To measure elapsed time, OSL relies on counting electrons trapped in crystal lattices since the grains were last on the surface. Entrapment occurs at a steady rate over time, per locality, which is why OSL can be used as a chronometer. Electrons are suddenly released, literally in a flash of light, when a sample is heated. Results do not require calibration and are quoted in calendar years. With suitably preserved material, the useful range of OSL is on the order of 1 million years. Its temporal reach is thus two orders of magnitude greater than radiocarbon's. OSL is just beginning to make a real impact on Quaternary paleontology, and its importance is sure to grow.

GLOSSARY

The following list includes terms and phrases used throughout this volume, for ready reference. Within an entry, **boldface** indicates another entry in this glossary.

Bering land bridge; Beringia: The Bering Strait, separating the Chukotka Peninsula in easternmost Russia from the Seward Peninsula in westernmost Alaska, is extremely shallow (its average depth is 30–50 m/100–150 ft). During periods of lowered **sea level**, such as the interval from about 75,000 to 20,000 years ago, the strait's seabed would have been widely exposed, forming at times an intercontinental land bridge. The hinterlands lying at either end of the land bridge, together with the bridge itself, constituted Beringia. (By convention, the Asian side of Beringia is called "western Beringia," the North American side "eastern Beringia.")

Biological first contact: Biological first contact is modeled on the concept of anthropological first contact, which marks the point in time when peoples of different cultural backgrounds first met and began to interact. Similarly, biological first contact occurs when people arrive in a place in which they have never been previously resident, and begin to interact with the local ecosystem and its biota. The anthropological sense of first contact also embraces the multiple consequences that follow along with such encounters. In human history, first contacts between empowered and powerless cultures have often meant dominance for the former and disaster for the latter. In the context of natural history, the result has been much the same, which makes the metaphor highly appropriate.

Body size: For extinct animals represented only by their skeletons, body size (weight) can be estimated by comparison to species of known size. Methods such as regression analysis take advantage of the fact that the dimensions of certain body parts (e.g., molar teeth, long bones) vary in a consistent manner with body size across a broad range of species. However, such methods generally assume that extinct species were proportioned the same as extant species within a group. If they weren't, body size predictions may be highly misleading. Since we rarely know much about actual proportions I prefer very conservative estimates for extinct species.

BP: Before present, in calendar years. (Original radiocarbon dates are separately designated as "radiocarbon years BP," calculated as years before 1950 CE.)

CE: Common era; equivalent to AD in older methods of numbering years.

Cenozoic: The geological era spanning the interval between 66 million years ago to the present.

Clovis: North American Archaic or Paleo-Indian archeological culture. Sites of Clovis activity, distributed throughout the continent south of ice sheets, date to between 13,200 and 12,900 years ago. The Clovis culture had a distinctive style of artifact manufacture, including production of fluted, bifacially flaked projectile points ("Clovis points"); the techniques involved were also developed by **Solutrean** peoples, evidently independently (but see Bradley and Stanford, 2004).

Dansgaard-Oeschger (D-O) event: Very rapid but short-lived climate events registered in Greenland ice cores, consistent with rapid warming followed by a slow decline back to previous (lower) temperature regime. The cause is uncertain.

Dreadful syncopation: In music, syncopation refers to stress on a normally unstressed beat. Off-beat stress results in "rhythmic surprise," because it is unexpected. In the context of the overkill of blitzkrieg hypothesis, the surprise was the devastating, recurring blow to biodiversity worldwide associated with the entry of humans into previously unvisited or unexploited environments.

First appearance date: As used here, first appearance dates refer to the earliest (usually archeological) evidence for the peopling of specific places (e.g., 1250 CE for the arrival of humans in New Zealand). This is a convention similar to that for **last appearance date**, and potentially open to same sources of error.

Fauna: A fauna is the totality of animal species living at one time in a defined area.

Genome: The entirety of the genetic material belonging to an organism. Its study or analysis is known as genomics.

Glaciations; interglaciations; stadials; interstadials: Conventionally, cold episodes are termed glaciations (or glacials) if long (around 100,000 years), stadials if much shorter (around 1000 years). Warm episodes are termed interglaciations (or interglacials) if longer than 10,000 years, interstadials if much shorter. "Warmer" and "cooler" are not absolute; they are just one-word summaries of "average" temperature relative to whatever went before. The Late Pleistocene was predominantly cold compared to today, so the "warmer" interstadials interspersed within that interval should not be thought of as being all that similar to the present.

Heinrich event: Short-lived climate events (usually lasting fewer than 1000 years) recorded in Pleistocene marine sediments and involving sudden cooling, presumably due to an influx of great quantities of ice-derived meltwater into North Atlantic from continental ice sheets. The cause is uncertain, but in theory large quantities of fresh water rapidly discharged into the Atlantic could disrupt oceanic thermohaline circulation, causing atmospheric temperature drops and affecting humidity in various ways worldwide (see Leydet et al., 2018).

Holocene: The most recent geological epoch, coextensive with the most recent (current) **interglaciation** beginning 11,700 years ago. At present, there are no accepted divisions into formal subunits; in future, the interval after 1945 CE may be separately designated as the Anthropocene, but this is not yet official.

Hominini: Name for the evolutionary group (technically a tribe) to which all species in the genus *Homo* belong, plus close relatives not discussed here (*Australopithecus* and related taxa).

Ice Age: Informal usage for the **Quaternary** when emphasizing the role of major climate change, to focus on its most obvious aspect: the formation, advance, retreat, and disappearance of ice sheets. Within this interval, there were many such advances and retreats of the ice, some of which have formal names and others of which are just squiggles within a larger field of squiggles on a chart.

Interglaciations: *See* **Glaciations; interglaciations; stadials; interstadials.**

Interstadials: *See* **Glaciations; interglaciations; stadials; interstadials.**

K/Pg: Abbreviation for Cretaceous/Paleogene boundary, which is currently placed at 66.02 million years ago. It is often used in reference to the mass extinction event which occurred at that time (in older terminology, called the K/T, or Cretaceous/Tertiary boundary).

Last appearance date: By convention, the "last" time that an extinct species was observed or detected, as an estimate for the actual date of final loss, usually unknown and unknowable. For most purposes, this is an acceptable procedure, especially if the estimate is based on a large number of radiocarbon dates (e.g., the last appearance date for Wrangel Island population of the woolly mammoth is 3730 ±40 radiocarbon years ago, or about 4000 to 4200 calendar years ago). For very recent extinctions, there is sometimes documentary evidence for a specific year (e.g., the documented last sighting of the Falkland Islands dog or warrah is 1876 CE), although once again the last recorded sighting is not necessarily of the last animal.

Last Glacial Maximum: The Late Pleistocene interval during which global ice volume was at its height. Although definitions and sources differ, the best current estimate for this interval is 27,500 to 23,300 years ago, which can be simplified to its midrange, i.e., 26,000 years ago. This was also the time during which the **sea level** was at its lowest and atmospheric dust at its highest level at any time in the past 130,000 years. After the maximum, it took many thousands of years for the ice sheets to retreat to their present limits. Also, the major continental ice sheets were not completely in phase, so the timing of local glacial maxima on different continents was not the same. This can be explained by the fact that, in addition to everything else, the Last Glacial Maximum was a time of extremely dry conditions, because cold air is less able to hold water vapor than warmer air. Thus, some precipitation-starved glaciers were actually in retreat during the Last Glacial Maximum.

Last Interglaciation: The last long warm interval before the present one (**Holocene**). Like **glaciations**, the onset of the Last Interglaciation and its temperature optimum varied across the planet. In the Northern Hemisphere, the current estimate for its length is 130,000 to 123,000 years ago. At least in midlatitude North America, pollen evidence suggests that the Last Interglaciation was as much as 3°C (5.4°F) warmer than today—a very significant amount in light of our current concerns about global warming.

Late Pleistocene: A formal geochronological unit (subepoch) corresponding to the last part of the Pleistocene epoch, 126,000 ±5000 years ago to 11,700 years ago (often rounded up to the interval between 130,000 and 12,000 years ago). As a formal unit, "Late" is capitalized (as are Early and Middle Pleistocene).

Late Quaternary: As used here, approximately the last 130,000 years, thus embracing the **Late Pleistocene** and **Holocene**. As an informal unit, "late" is not capitalized in this term.

Modern era: The last 500 years (1500 CE to present), an interval marked by the greatest number of new **biological first contacts** and subsequent extinctions since the diasporas of Near Time.

Near Time extinctions: Species losses occurring between 50,000 and 500 years ago. (Losses in the past 500 years are grouped as modern era extinctions.) Near Time is capitalized for emphasis, but it is not a formally defined interval like a geological epoch.

Pleistocene–Holocene boundary: The temporal boundary dividing the Pleistocene from the Holocene was previously defined as an exact point in time—specifically, 10,000 calendar years ago. This is no longer accepted. It has been redefined to coincide with a major climate shift detected in an ice core collected by the North Greenland Ice Core Project (NGRIP) on the Greenland ice sheet. Annular ice layers entrap chemical and particle (dust) signatures for both atmosphere and precipitation at the time of their deposition, which provide information on climate. In the NGRIP core, abrupt changes in various parameters can be detected at a position within the core corresponding to 11,700 calendar years ago, with a maximum layer counting error of ±99 calendar years. In combination, these markers define the end of the Younger Dryas cold spell and the coincident beginning of the Holocene.

Pleistocene–Holocene transition: The transition between these epochs is a much wider interval than the **Pleistocene–Holocene boundary**. Here, it will be considered to represent the interval between 14,500 and 9,000 years ago, which incorporates several warming or cooling events that have been correlated with the decline and eventual disappearance of many vertebrate species worldwide.

Pliocene: The geological epoch immediately before the Pleistocene, running from 5.3 to 2.6 million years ago. For this book, its chief importance is that the first of the major ice sheet advances took place near the end of the epoch.

Prey naïveté: In behavioral terms, the failure of prey animals to recognize and respond to predators, with the predictable consequence that they experience enhanced predation.

Proxy: In scientific usage, a proxy is usually something measurable that varies in roughly the same way as something that cannot be measured directly, and thus can stand in for it. For example, scientists can infer ancient air temperatures from oxygen isotopic concentrations recovered from ice cores. It is important to emphasize that the connection between a proxy and an inference based on it necessarily involves assumptions that may not apply in all circumstances.

Quaternary: The most recent period of Earth history, the last 2.6 million years, divided into two epochs, the Pleistocene (2.6 million to 11,700 years ago) and the Holocene (11,700 years ago to the present). As formal geochronological terms, they are always capitalized.

Sea level: Scientists employ special stratigraphic methods to establish past changes in sea level. The highest sea level during the late Quaternary occurred around 120,000 years ago: 6 to 9 m (20 to 30 ft) above today's sea level. The lowest occurred during the late stage of the Last Glacial Maximum, around 23,000 to 25,000 years ago, when the sea level sank as much as 135 m (440 ft) below modern levels.

Solutrean: Late Paleolithic archeological culture, represented in sites in western Europe dated to between 22,000 and 17,000 years ago. Production of Solutrean stone artifacts involved several innovations not in evidence in earlier industries, such as use of soft hammers in bifacial percussion techniques.

Taphonomy: The study of how organisms decay and become fossils.

Thermohaline circulation: Think of the world ocean being slowly moved around on a series of interconnecting conveyer belts that transfer water from one place to another, both latitudinally and vertically. The belts are largely driven by differences in temperature and salinity (hence thermohaline) in the water column, one result of which is the gradual movement and global mixing of oceanic waters over time. In the Atlantic, the transport of warm water from the tropics northward via the Gulf Stream moderates higher-latitude temperatures over both land and sea. Conveyers in other parts of the world ocean function similarly, moving heat energy and inducing mixing across great distances.

Disruptions of the conveyor belt system are thought to have happened repeatedly during the **Quaternary**, affecting climates globally (see **Heinrich event** and **Dansgaard-Oeschger [D-O] event**).

Wisconsinan glacial episode: The last **glaciation**, formally defined as running from 85,000 to 11,700 years ago, corresponding to the Weichselian glacial episode in Eurasia. The amount of land surface overlain by ice varied considerably during this interval, but the greatest coverage occurred during the **Last Glacial Maximum**, centered on ca. 26,000 years ago. By 20,000 BP the ice was in retreat in the Northern Hemisphere. According to current information, retreat in the Southern Hemisphere was delayed until ca. 12,500 years ago.

Younger Dryas: A brief but intensely cold period between warmer intervals near the end of the Pleistocene, 12,900 to 11,700 years ago. Start and end dates are based on stable isotope data from Greenland ice cores, which record these dramatic shifts. Preceded by the Allerød **interstadial**, the Younger Dryas was followed by the Holocene **interglacial**.

NOTES

1. Big

6 **When people hear:** Foodies and perhaps others may enjoy reading whether mammoth was ever served up, as urban legend has it, at the Explorers Club in New York City. It wasn't: see MacDonald (2016).

8 **Take the duo dominating:** The Columbian mammoth (*Mammuthus columbianus*) is usually regarded as a species separate from the woolly mammoth, although recent genetic evidence shows that they were capable of hybridizing (see Enk et al., 2016). In any case, both types of mammoths appear to have died out at the same time, 12,000 to 13,000 years ago, except for small refugial populations.

2. "This Sudden Dying Out"

14 **I have no objection to this usage:** Regarding the concept that there are always both winners and losers in mass extinctions, see Jablonski (2001).

15 **low reproductive rates and slow maturation:** See Johnson (2002).

15 **In North America at present:** The number of species-level extinctions in the continental New World varies, depending on species definitions. I am admittedly conservative.

19 **Although many theories:** Whether passenger pigeons experienced dramatic fluctuations in population sizes and genetic diversity in the period leading up to their disappearance is the subject of considerable disagreement. According to Murray et al. (2017), pigeon populations appear to have actually been relatively stable, but were not genetically diverse for adaptational reasons. This situation may have contributed to the severe impact of overhunting in the nineteenth century, as the species lacked sufficient genetic resources to deal with the level of persecution they experienced. On this argument, rapid collapse was inevitable. Lesson: having large and stable population sizes is not always a buffer against extinction.

19 **Number of passenger pigeons:** Audubon (1832, 1:322).

21 **Beginning in the 1960s:** Martin's mature summary of his views is contained in his last book (Martin, 2005).

24 **However, I think a plausible estimate:** See lists assembled by Turvey (2009a). See especially Sandom et al. (2014) for a recent global assessment of losses in all terrestrial vertebrate groups.

3. The World Before Us

28 **The epoch immediately preceding:** See Funder et al. (2001).

28 **Many causes for the changeover:** This topic is too complex to cover here. A very readable account of Milankovitch cycles is provided at https://en.wikipedia.org/wiki/Milankovitch_cycles.

30 **In the coldest part of the last glaciation:** This section on Quaternary environments and geology is mainly based on relevant chapters in Ehlers, Hughes, and Gibbard (2016).

35 **Although Late Pleistocene conditions:** The highly diverse meridiungulates, a group probably unfamiliar to most readers, were South American placental mammals distantly related to horses, tapirs, and rhinos (order Perissodactyla). By the Late Pleistocene, they had been reduced to a handful of species (e.g., plate 11.1). The gomphotheres represented a third family of late Cenozoic proboscideans in addition to mammoths (Elephantidae) and mastodons (Mammutidae). Gomphotheres were the only proboscideans that managed to get to South America, where they lasted until the end of the Pleistocene (see plate 8.3).

37 **Southern California panorama:** An up-to-date list of La Brea fossils (vertebrates as well as some invertebrates) can be found at https://en.wikipedia.org/wiki/List_of_fossil_species_in_the_La_Brea_Tar_Pits.

40 **The art clearly depicts:** The Bradshaw Foundation maintains an archive of African rock art at http://www.bradshawfoundation.com/africa/index.php.

4. The Hominin Diaspora

44 **If these Javanese hominins:** See van den Bergh et al. (1996a).

44 **Anatomically modern people:** See Hublin et al. (2017).

48 **From about 120,000 years onward:** See Bae, Douka, and Petraglia (2017); Westaway et al. (2017).

48 ***H. sapiens* finally made the crossing:** See Hiscock (2008).

48 **We will focus on recent work:** For reflections on the debate over initial human entry into the New World, from an archeologist who has been in the thick of it for most of his career, see Meltzer (2010, 2015).

48 **The corridor permitted:** Heintzman and colleagues (2016) show how fluctuations in genetic continuity among bison populations located on either side of the major North American ice sheets can be used as a proxy for determining when the corridor was open and available for human migration.

49 **Recent studies have fixed the last period:** See Pedersen et al. (2016). See also Heintzman et al. (2016).

49 **the Page-Ladson mastodon kill site:** See Halligan, Waters, and Perrotti (2016).

50 **Although stone artifacts have not been recovered:** See Bourgeon, Burke, and Higham (2017).

51 **There are other theories:** See Bradley and Stanford (2004).

51 **Moreover, this hypothesis has come up against:** See Llamas et al (2016).

52 **Steve Holen and Tom Deméré:** See Holen et al. (2017).

53 **The oldest dated human remains:** See Chatters et al. (2014).

53 **The site of Monte Verde:** See Borrero (2009); Fiedel (2009); Haynes (2009a).

53 **This was controversial enough:** See Dillehay et al. (2015); Braje et al. (2017). Fariña et al. (2013) make a hesitant claim for an even earlier presence of humans—30,000 years ago—in South America, the evidence in this case being possible cut marks on limb bones of the giant sloth *Lestodon* and other megafauna at the site of Arroyo del Vizcaíno, Uruguay. There is also a single possible tool identified as a scraper. They do allow, however, that the cut marks may be no more than "an example of natural processes mimicking human presence."

53 **the humans would have been occupying:** Human occupation of the high Andes may be much older than previously thought, for there is some latest Pleistocene evidence for occasional human presence in the Bolivian Andes. See Capriles et al. (2016); Haas et al. (2017). By 7000 BP human occupation was permanent, in the sense that people were living full-time at 3000 m (10,000 ft) instead of migrating back and forth from the surrounding lowlands over the course of the year. As this was well before

any evidence of agriculture, the occupants must have followed a hunting and gathering way of life. Whether humans were hunting megafauna has not yet been established.

53 **Apart from specialist disagreements:** After Monte Verde, the next oldest occupation site in South America may be the site of Arroyo Seco 2, dated to 14,000 years ago. This site, which includes lithic artifacts and the remains of extinct mammals (sloths and horses), is in the Pampas region of Argentina, on the opposite side of the Andes from Monte Verde and in a quite different ecological setting. See Politis et al. (2016).

55 **Malta panorama:** For body sizes and stature, see van der Geer et al. (2014, 2016).

56 **On most islands:** See van der Geer et al. (2010).

56 **Alan Simmons of the University of Nevada:** Simmons (1999) states his case for why humans were probably responsible for the concentration of hippo bones, but acknowledges that the complete absence of cut marks is difficult to account for.

56 **A number of extinctions occurred:** See MacPhee (2009); Cooke et al (2017).

60 **the world's fourth largest island:** See Dewar et al. (2013).

60 **What we do know:** On cut marks on Malagasy hippo bones, see MacPhee and Burney (1991). More recently, cut marks have been identified on sloth lemur bones, also dated to about 2000 years ago (see Godfrey and Jungers, 2003).

60 **They may have comprised:** See Holdaway et al (2014); Allentoft et al. (2014).

60 **no other islands have yielded:** Bone-bed sites have been discovered in New Zealand (e.g., at the mouth of the Shag River, South Island) from which many thousands of bones have been recovered in the aggregate. However, most New Zealand sites are much less than a millennium old, and thus do not compare to end-Pleistocene localities in North America with regard to loss of material through weathering and degradation. See papers edited by Anderson, Allingham, and Smith (1996).

63 **Canterbury Plains panorama:** see Worthy and Holdaway (2002).

63 **"such bulky and probably stupid birds":** see Owen (1844, 73).

64 **Researchers Michael Carleton and Storrs Olson:** See Carleton and Olson (1999).

64 **And so it goes with island extinctions:** Humans have the distinction of being the first Quaternary mammals other than bats to invade New Zealand, but they had company. Traveling with the ancestors of the Maoris to New Zealand was a particularly resourceful rodent, the Polynesian or kiore rat (*Rattus exulans*). The rat rapidly achieved a wide distribution on North and South Islands, where it seems to have caused the local extirpation or the outright extinction of a variety of small vertebrates and invertebrates, including lizards, flightless beetles, and land snails. Direct evidence for this is essentially absent, but the scenario is plausible. Although sometimes thought of as relying mainly on plant parts and seeds, many members of the mouse/rat family can easily shift to an omnivorous diet. See https://en.wikipedia.org/wiki/Polynesian_rat.

5. Explaining Near Time Extinctions: First Attempts

68 **Those interested in more detail:** A useful, in-depth consideration of how extinction was understood and explained by nineteenth-century authors can be found in Grayson (1984). See also Rudwick (1976).

68 **"Life, therefore, has been often disturbed":** Cuvier (1829, 11).

69 **ancient peoples must have come across megafaunal fossils:** The Wikipedia webpage on Cyclops harrumphs that the story of its monocular head having been confounded by the ancients with the skull of a dwarf elephant is a "modern myth" because none of the classical sources make any sort of reference to "skulls, or cyclops, or even elephants, which were unknown to Greeks at the time" (https://en.wikipedia.org/wiki/Cyclops). Of course, it is just a fantasy. On the other hand, as Adri-

enne Mayor (2000, 2005) shows, fossils of many different kinds were known to the ancient Greeks and other peoples, whether or not they knew what they were looking at. It is certainly not beyond all likelihood that adventurous spelunkers in classical times might have chanced upon the remains of dwarf elephants and other Quaternary fossil vertebrates, and wondered. . . .

72 **his great calamities:** See Cuvier (1829, 11). Cuvier believed that mammoths died out during the "last" revolution, when "the surface of our globe ha[d] been subjected to a vast and sudden revolution, not further back than from five to six thousand years" (1829, 181). Remains of both mastodons and mammoths had been discovered in North America during the eighteenth century, but confusion reigned as to what, exactly, they were and their relationship to extant elephants. See Semonin (2000); Dugatkin (2009, 83).

72 **Religiously inclined scholars:** Cuvier was cagey about the Flood and human antiquity. He reasoned that, although no human bones had ever been discovered in clear association with the remains of extinct animals, if humans did exist back then then perhaps they managed to avoid catastrophe in "some confined tract of country, whence [they] re-peopled the world after the terrible events" of the Flood (1829, 85).

72 **Louis Agassiz's investigation:** Agassiz's concept of the Ice Age was developed in the 1830s. The quotation in this paragraph is from a later compendium written for an English-language audience (Agassiz, 1866, 208).

72 **It pointed to a past catastrophe:** However, as an anti-Darwinist, Agassiz was strongly against adaptive evolution, as then understood, and made other implausible claims, including one to the effect that ice sheets had covered all of North America, pinching out life everywhere on the continent. The biota of the present epoch was therefore new life, specially created, not the result of repopulation by Ice Age survivors.

73 **When fossils of the American mastodon:** See Semonin (2000); Dugatkin (2009). Quotations are from a contemporary broadsheet and Rembrandt Peale's statement defending his reconstruction, reproduced by Semonin (2000, 329, 335).

75 **"some great system":** The entire quotation, from p. 53 of Darwin's holograph Notebook B, reads: "Whether extinction of great S. American quadrupeds [was] part of some great system acting over whole world[,] the period of great [Pleistocene] quadrupeds declining as great reptiles must have once declined." Found at http://darwin-online.org.uk/content/frameset?keywords=cul%20dar121&pageseq=1&itemID=CUL-DAR121.-&viewtype=side.

75 **"We may presume that the time demanded":** Lyell (1866, 374).

76 **"vain to contend":** Buckland (1831, 610).

77 **"It is probable that causes more general":** Lyell (1866, 374).

77 **"We need not marvel":** Quoted in Quammen (2008, 325).

77 **"We cannot but believe":** Wallace (1876, 150). Wallace's contribution to identifying the cause of Ice Age extinctions was limited, but he did notice something important about how size affected fecundity:

> There is, however, another cause for the extinction of large rather than small animals whenever an important change of conditions occurs . . . but which has not, I believe, been adduced by Mr. Darwin or by any other writer on the subject. It is dependent on the fact that large animals as compared with small ones are almost invariably slow breeders.

Here we have an inkling of an insight into one of the major pattern features of the megafaunal extinctions: physiology, not size as such, is what matters because it correlates with reproduction rate. As was typical for Wallace, he thanked a "correspondent," Mr. John Hickman of Desborough, otherwise unknown, for giving him this idea. Interestingly, Osborn (1906, 852) huffily dismissed Wallace's slow breeder argument, stating that "it receives no support from paleontology."

78 **"[The] overwhelming trend":** See Romer (1933, 76–77).

78 **"[I]t may have been the entrance":** Osborn (1910, 507).

6. Paul Martin and the Planet of Doom: Overkill Ascendant

81 **"one agency that might have produced":** Holder (1886, 47–48).

82 **"within one or two score thousands of years":** Romer (1933, 77).

82 **Paul Martin's contribution was:** See any of his influential papers on the importance of accurate radiocarbon dating for testing overkill, especially Martin (1973, 1984, 2005). At a conference held at the American Museum of Natural History in April 1997, aptly entitled "Humans and Other Catastrophes: A New Look at Extinction and the Extinction Process," I recall his stating in his lecture that, without radiocarbon dating, there never would have been a debate over the cause of Pleistocene extinctions. What he meant was that, absent any ability to accurately date first appearances of people and last appearances of megafauna, discussion of how these extinctions might have occurred would have probably remained at the stage of hazy assertion exemplified by the works of earlier writers. Accurate dating is the indispensable first step to making sense of Near Time extinctions (cf. Zazula et al., 2014).

82 **For example, some early authors:** Humbert's (1927) hypothesis, that the depauperate grasslands of most of central Madagascar were due to ecological change caused by unrestricted burning when people first arrived, was generally accepted into the 1980s. David Burney showed in a series of papers (e.g., Burney, 1993) that the grasslands were much older than the late Holocene arrival date of humans. This is not to say that burning is irrelevant to species endangerment in Madagascar at present (see Goodman and Jungers, 2014); whether it was contributory to subfossil extinctions is unresolved.

83 **Initially, Vartanyan's radiocarbon results:** The title of the initial publication in *Nature* (see Vartanyan, Garutt, and Sher, 1993) strongly implied that the Wrangel Island mammoths exhibited reduced body size, but this was incorrect. However, they had lots of other problems, including inbreeding and associated genetic disasters (see Rodgers and Slatkin, 2016).

85 **As the game is played out:** Cheetahs hunting impala are successful in bringing down their targets about one time in three, in part because prey adjust their speed in accordance with that of the predator to save energy for final escape maneuvers. See Wilson et al. (2018).

85 **The scales might tip:** A long-term study of wolves and moose on Isle Royale in Lake Superior considers consequences of loss of equilibrium in predator-prey relationship; see Mlot (2017).

87 **The most arresting evidence:** Excellent, well-illustrated sources on Paleolithic art are Bahn and Vertut (1997); Guthrie (2005).

88 **The most desirable prey species:** Although overkill arguments tend to focus on meat, for basic nutritional reasons fats would have been of equal or greater importance to early paleohunters. In the view of Ran Barkai of Tel Aviv University, acquisition of high-quality fats supplies a possible reason that mammoths were favored targets. Humans do not have guts that can efficiently extract protein from meat. Fat, on the other hand, can be consumed immediately and without limit, and it is a higher calorie source than meat (fat yields 9 calories/gm, meat only 4 calories/gm). See Zutovski and Barkai (2016).

89 **By the start of the Holocene:** Modern bison narrowly avoided complete extinction at the end of the nineteenth century, thanks to massive efforts to save the last members of the species. How many individuals were left by that time is uncertain; guesses range from a few hundred to less than a thousand. By contrast, estimates for the number of bison just before European arrival in the New World run

from 35 million on the low end to 65 million on the high. That means that the bison population in 1880 was probably about 0.002 to 0.003 percent of what it had been in 1500. That bison could be brought back at all after losing much more than 99.9 percent of the standing crop in the course of a few centuries is remarkable. Bison were certainly hunted by early humans, but because of rampant taxonomic splitting it is not clear whether any truly distinct species disappeared at the end of the Pleistocene. For more details, see Nowak (1999, 2:1161).

7. Action and Reaction

92 **"To single out a particular predator":** Guilday (1967, 121).

93 **"there are no modern analogues":** Graham and Lundelius (1984); see also Graham (1985).

97 **The appeal to ecological theory:** Another ecologically based argument that was influential at this time was the "keystone herbivore" hypothesis, mainly associated with the work of Norman Owen-Smith (see Owen-Smith, 1999). Keystone species are ones that effectively create habitat for other species, and in that sense the hypothesis is coevolutionary. In Africa today, elephants and rhinos keep forests open by bulldozing through the brush, thereby creating trails and clearings in which smaller animals can forage. Mammoths, mastodons, ground sloths, and notoungulates like *Toxodon* may have provided a similar service, and with their extinction these supermegaherbivore roles were lost. Whether their disappearance prompted complete extinctions of other species that relied on them, however, is a different matter.

102 **Martin, always interested:** For a discussion of his views on the inadequacies of climate-based arguments, see Martin (1984) and Martin and Steadman (1999).

103 **In sum, Martin was looking:** Regarding the phrase "dreadful syncopation," see the original definition by MacPhee and Marx (1997) and Martin's (2005) later incorporation of it into the language of blitzkrieg.

103 **the elimination of woolly mammoths:** See Nogués-Bravo et al (2008).

8. Overkill Now

108 **Paul Martin's overkill hypothesis:** See Horan, Shogren, and Bulte (2003); Russell (1995).

108 **Now past genetic diversity:** For hybridization and "invisible extirpation" among mammoths, see Enk et al. (2011, 2016).

109 **the extinction window could have been as narrow:** See Fiedel (2009).

109 **Among the most influential of these:** See Alroy (1999, 2001); Smith et al. (2018).

109 **the climate-versus-people debate:** For additional evaluation of Alroy's model, see Brook and Bowman (2002).

114 **controversial evidence for the apparent persistence:** For example, see Haile et al (2009); Barnosky and Lindsey (2010).

114 **Speaking of sloths:** See Steadman et al. (2005). Also see MacPhee, Iturralde-Vinent, and Jiménez-Vázquez (2007) on additional evidence for late sloth extinction in Cuba.

114 **the primary lesson of the Antillean extinction story:** See Martin (2005); see also MacPhee (2009).

114 **Early hominins:** In Europe hominins were hunting straight-tusked elephants as early as Middle Pleistocene at the site of Marathousa, Greece (see Panagopoulou et al., 2015).

118 **The bear was a thoroughgoing herbivore:** See Naito et al. (2016).

120 **Did they die out:** Although some now extinct megafaunal species did manage to survive into the Holocene, it is unlikely that the number was large. Turvey et al. (2013) have shown, for example, that a number of alleged megafaunal survivals into the Holocene in China (e.g., woolly rhino) are based on poor or suspect evidence.

120 **the persistence of woolly mammoths:** For last dated survival of mammoths on mainland Eurasia, see MacPhee et al. (2002).

120 **about a millennium longer:** See Pitulko et al. (2016).

120 **This is higher than the loss factor:** On loss estimates for different areas of the globe, see Wroe et al. (2004).

120 **Recently, a research team:** See Clarkson et al. (2017).

121 **In recent years, an unusual proxy:** See Burney, Robinson, and Burney (2003).

121 **Fairly consistent results:** See Burney, Robinson, and Burney (2003); Feranec et al. (2011); Rule et al. (2012).

124 **With the exception of fossils:** On likelihood that Cuddie Springs is disturbed, see Wroe et al. (2004).

124 **On the basis of extensive excavations:** See Cosgrove et al. (2010); Cosgrove and Garvey (2017).

125 **Ninja!:** See White et al. (2010).

127 **However, Tim Flannery:** See Flannery (1995).

127 **While this scenario differs:** Whether or not Flannery's view concerning prehistoric Sahul is correct, on a global basis anthropogenic environmental transformation started far back in antiquity. For example, there is evidence that humans have been substantially altering forests in the tropics for at least 45,000 years (see Roberts et al., 2017).

128 **Their simulations indicated:** See Saltré, Johnson, and Bradshaw (2016). For a balanced summary of the possible human role in the Australian extinctions, see McGlone (2012).

129 **In sharp contrast to continuing disagreement:** Worldwide losses among placental mammals on islands during the late Quaternary are exhaustively catalogued by van der Geer et al. (2010). For the modern era (the last 500 years), also see MacPhee and Flemming (1999).

129 **Thus in the case of Antillea:** See Cooke et al. (2017). Introduced species are powerful agents of extinction, especially on islands. In recent years, brown tree snakes (*Boiga irregularis*), introduced on Guam, have arguably caused the local extirpation of as many as twelve bird species, and black rats have long been thought to have caused the loss of small vertebrates on numerous island in Antillea 500 years ago.

132 **Javanese dwarf stegodon and water buffalo:** On dwarf stegodons of Indonesian islands, see Van den Bergh et al. (1996b).

133 **Recent bird extinctions throughout the Pacific:** See Steadman (2006).

133 **At the site of Liang Bua:** See Gramling (2016). Modern *Homo sapiens* was surely present on many Indonesian islands by 50,000 years ago, although the earliest physical evidence of anatomically modern humans on Flores is only latest Pleistocene.

9. Where Are the Bodies, and Other Objections to Overkill

135 **"if these bones should prove":** Hyslop (1988, 153).

136 **"An explosive model":** Martin (1973, 972).

136 **the occasional Clovis point:** On this issue, see Meltzer's (2015) discussion of possible reasons that direct evidence is so rarely encountered.

139 **Among the few megamammals:** See Borrero (2009); Fariña et al. (2013); Dillehay et al. (2015); Politis et al. (2016).

139 **Gary Haynes, a professor:** See Haynes (2002, 2009a).

139 **We assume that nonavian dinosaurs:** The Chicxulub bolide 66 million years ago was an equal-opportunity impactor: body masses meant nothing. Theropod dinosaurs as small as half a pound (250 gm) died out just the same as 10 ton tyrannosaurs. Yet the patterning of end-Cretaceous losses still eludes explanation in many respects. For example, there was differential survival among the

microorganisms that populate the seas worldwide. Virtually all lineages of calcareous nannoplankton, which live in surface waters, suffered massive losses (more than 99 percent of species then living). By contrast, other organisms such as benthic (deep-sea) foraminifers mostly survived, implying that the great thickness of the water column acted as a buffer, diluting the deleterious effects of the impactor. But there were no environmental protections for nonavian dinosaurs; they all perished, despite the fact that they lived in exactly the same places as true birds, who survived.

139 **the number of good dating records:** See Grayson and Meltzer (2003, 2015).

139 **For example, extinct species of camels:** See Waters et al. (2015).

140 **Although there is something to be said:** Generalizing about the effect of a preferred driver is a common practice where large-scale extinction events are concerned, given that it would frankly be an impossible job to produce an independent last appearance date for every species presumed to have disappeared at a given time.

140 **were the hunters performing:** Horan, Shogren, and Bulte (2003) model the consequences of overkill and aspects of coevolution from an economics viewpoint.

141 **In his later work:** See Martin and Steadman (1999).

141 **If this last comes to pass:** It may have already come to pass in the case of the northern white rhino; no members of this subspecies have been seen since 2006–2007, and it is "possibly extinct in the wild" according to International Union for Conservation of Nature criteria. See http://www.iucnredlist.org/details/4183/0.

141 **Like poachers today:** See papers, variously pro and con, by Grayson and Meltzer (2015); Waguespack and Surovell (2003); Lyons, Smith, and Brown (2004a).

141 **In Africa, this kind of market hunting:** See Bobo et al. (2015).

142 **Hunting Merck's rhinoceros:** Starkovich and Conrad (2015) assess the composition of faunal remains from Schoneningen, which includes *Stephanorhinus*. This site is 380,000 to 400,000 years old, which underlines the point that rhino hunting extended over a long period, involving hominids other than *Homo sapiens*.

143 **Some other late Quaternary kill sites:** See Faith and Surovell (2009).

143 **losses were sudden rather than staggered:** But see Haile et al. (2009) for evidence of late survival of mammoths and horses in interior Alaska.

143 **Archeologist George C. Frison:** See Frison (1998). For those interested in the definition of "bands" in ethnography, I recommend Ingold's (1999) treatment.

144 **As to meat caching:** Anthropologists and archeologists sometimes make a distinction between forager and collector strategies in hunting and gathering. Essentially, foragers are always on the move, exploiting resources where they can. Collectors by contrast send out specialized parties to recover targeted resources that are then brought back to a more or less stable campsite. Allowing that such activities are on a spectrum, collectors, but not foragers, are in a position to acquire surpluses (see Bettinger, 1987). Considering the bounty of megaherbivores in Late Pleistocene North America, it would make sense if some paleohunter groups became collectors. But that would be true for "normal" hunting (and scavenging); unrestricted hunting as envisaged by Martin's bow-wave hypothesis is a very different undertaking. If collecting and storage/distribution at home bases was a key strategy for ancient North American big-game hunters, it becomes even harder to understand how they could have simultaneously acted as extinction agents over wide areas, preventing any species recovery.

146 **This flies in the face:** See Gowdy (1999). A brief topical introduction to hunter-gatherer ethnology and its recent literature can be found in Ember (2014).

146 **Killing prey over and above:** See Stuart (1986); DelGiudice (1998).

146 **"[s]atiation in carnivores":** Kruuk (1972, 233).

147 **a study on introduced predators:** See Short, Kinnear, and Robley (2002).

10. More Objections: Betrayal from Within?

149 **Great and small, moa and wren:** See Millener (1988).

150 **Martin's conception of faunal blitzkrieg:** See Martin (1973; 2005, ch. 10).

150 **"the population of vulnerable large animals":** Martin (1973, 970).

152 **Martin's implicit requirement:** On unlikelihood of a uniform failure to recover, see MacPhee and Marx (1997).

152 **Why didn't the western camel:** Yukon paleontologist Grant Zazula has drawn attention to the difficulties species in the midst of population collapse would have in retaining or regaining range. After 15,000 BP but before the onset of the Younger Dryas cold snap around 12,900 years ago, climatic conditions were mostly warmer. As the ice sheets began to withdraw, American mastodons, Jefferson's ground sloth, and western camels should have been able to return to the far north, as they had during the Last Interglaciation. For example, the postglacial boreal forests that rapidly established themselves across Canada at this time should have been prime territory for mastodons, enabling their reoccupation of parts of high-latitude North America previously covered by ice. Yet there is no evidence this happened, not in the case of mastodons nor that of any of the other large mammals which had surged northward during previous warming intervals. Were they already so depleted in numbers that reestablishment was impossible? See Zazula et al. (2014, 2017).

153 **Tarpan or wild horse:** See Nowak (1999, 2:1008ff).

155 **Indeed, as David Quammen remarks:** See Quammen (1996).

155 **"[These] pigeons":** Balouet and Alibert (1990, 79).

156 **"These wolves are well known":** See Darwin (1839, 194). Although there are some early statements to the effect that the warrah could act aggressively, most firsthand accounts resemble Darwin's.

156 **"island syndrome":** See Sánchez-Villagra, Geiger, and Schneider (2016).

156 **Paleontological evidence shows that features:** For several examples, see van der Geer et al. (2016; also personal communication). For instance, the Cyprus dwarf elephant *Palaeoloxodon cypriotes* was only 200 kg (440 lb; shoulder height 1.4 m/4.5 ft). Maltese *P. falconeri* may have been even smaller at 100 kg (220 lb; shoulder height 0.9 m/3 ft)—just 2 percent the size of its probable ancestor *P. antiquus*, one of the largest true elephants ever to have existed.

158 **Dwarf Cyprus elephant:** For the "fast" life hypothesis, see Raia, Barbera, and Conte (2003).

11. Other Ideas: The Search Continues

161 **"perhaps the strangest animal ever discovered":** Darwin (1839, 94).

161 **"greatly struck":** Darwin's Notebook B [On transmutation of species], found at http://darwin-online.org.uk/content/frameset?keywords=cul%20dar121&pageseq=1&itemID=CUL-DAR121.-&viewtype=side. See also https://en.wikipedia.org/wiki/Inception_of_Darwin's_theory.

162 **How many unique ecological:** See Marlow (2002) for a fascinating account of "gaps in nature" brought about by the loss of megafauna.

164 **Body size is one:** See Segura, Fariña, and Arim (2016).

166 **Although modern ecological research:** The elephants of Kenya's Tsavo region, faced with prolonged drought and wholesale starvation in the early 1970s, turned not to scavenging meat but, in the end, to tearing the bark off trees for what little nutrition it might provide. Female elephants and calves suffered the most because they tended to stay put. Bulls did better because they could travel to less affected regions. The elephants died in droves: in all, 25 percent of the Tsavo population was lost over a four-year period (see Corfield, 1973). Isotopic investigations of a number of Eurasian megafaunal species of Middle and Late Pleistocene age indicates more, rather than less, flexibility in both

diet and habitat, although whether this capacity was reduced in Near Time needs more study (see Pushkina, Bocherens, and Ziegler, 2014).

166 **collapse of species occupying one part:** See Whitney-Smith (2009).

168 **The disease hypothesis:** See MacPhee and Marx (1997).

170 **almost unbelievable levels of mortality:** For example, transmissible facial tumor disease in Tasmanian devils (*Sarcophilus harrisii*), or chytrid fungal infections in amphibians worldwide; see MacPhee and Greenwood (2013).

170 **The major weakness:** See Lyons et al. (2004b).

171 **However, in one instance:** See Wyatt et al. (2008). The final extinction of the thylacine or marsupial "tiger" may also have been due to reduced genetic fitness and consequent susceptibility to disease, as long suspected (see MacPhee and Marx, 1997). See Mao (2017).

172 **a 10-km-wide bolide:** See Firestone et al. (2007); Kennett et al. (2015); Hagstrum et al. (2017).

172 **In recent papers, fireball supporter:** See Kennett et al. (2015).

174 **Mucuñuque, in northwestern Venezuela:** See Mahaney et al. (2011).

173 **That particular kill mechanism:** The unlikely airburst argument has been revived and extended in order to explain why carcasses, bones, and even trees in Beringian "muck" sediments are "excessively damaged," as though torn apart by shrapnel (see Hagstrum et al., 2017).

174 **Recent investigations do support:** See Leydet et al. (2018).

175 **Why things should have happened:** See Cooper et al. (2015, 606); also see especially Metcalf et al. (2016). In South America, a warming phase began after 12,600 years ago, extending into the early Holocene. Since there is a concentration of megafaunal last appearance dates in Patagonia around 12,300 years ago, the authors make the point that the terminal die-off occurred at a time when the southern cone had been warming for several hundred years. At this point cold-adapted populations would have been stressed by the warming climate and its effects on vegetation (e.g., allowing forest expansion into grasslands). In a familiar scenario, they argue that that the final blow was human arrival. Things now took a quick turn for the worst, as the population seesaw went down and stayed down. The hypothesis that regular ups and downs in megafaunal population sizes due to climate change could have been disastrously interrupted by the advent of humans is interesting, but timing remains a key issue. For example, humans were already present in southern South America 14,500 years ago, if not considerably earlier, at the Chilean site of Monte Verde (see chapter 5). At that time, temperate South America would have been in the throes of a notably cool interval, known as the Antarctic Cold Reversal. If the humans were already taking megafauna at that time, did it take a further 2000 years, until the onset of the subsequent warming interval, before some combination of environmental change and human persecution finally produced its mortal effect?

12. Extinction Matters

179 **The "Sixth Mass Extinction":** Kolbert (2014) provides a good account of both the wellsprings of the current extinction crisis and its possible future outcomes.

180 **Not many archeologists think it probable:** See Braje et al. (2017).

181 **Reptiles and amphibians:** See Slavenko et al. (2016).

182 **A few other megafauna:** See Campos et al. (2010).

182 **small groups of hunters:** See Brook and Bowman (2002); Brook et al. (2013).

183 **modeling suggests that some prehistoric extinctions:** For a recent quantitative effort to assess the differential roles of climate change and human impacts, see Marshall et al. (2015). The authors emphasize that, to distinguish the contributory roles of different extinction drivers with some confi-

dence, the single greatest data need is accurate, precise dating of human arrival (i.e., first biological contact) in places where Near Time losses occurred.

184 **Fijian crocodile ambushes a megapode:** See Naish (2009). See also Worthy and Holdaway (2002).

Epilogue: Can the Megafauna Live Again?

187 **a series of Discovery Channel specials:** Disclosure: I participated in three of them: *Land of the Mammoth* (2001), *What Killed the Megabeasts* (2002), and *The Baby Mammoth* (2007).

187 **synthetic biology and genetic engineering:** For a very readable account of current de-extinction efforts, not limited to mammoths, see Shapiro (2016).

189 **no newborn mammophants:** George Church and others, including me, discussed the possibilities and problems of de-extinction at the 2017 Isaac Asimov Memorial Debate: see https://www.youtube.com/watch?v=_LnAtMeSVeY.

189 **Other scientists are aiming:** Successful progress in acquiring the genome of the passenger pigeon was recently announced; see Murray et al. (2017).

189 **bringing back the American chestnut tree:** See Powell (2016).

189 **This kind of "facilitated adaptation":** Among many other possibilities, there is the idea that we can save species by reengineering them; see Thomas et al. (2013).

189 **"rewilding":** See Martin (2005, ch. 10).

189 **Researcher Sergey Zimov:** See Zimov et al. (1995); Zimov (2005). An excellent documentary on Zimov's concept and its scientific basis is accessible through the Pleistocene Park website, http://www.pleistocenepark.ru/en/.

REFERENCES CITED

Agassiz, L. 1866. *Geological Sketches*. Boston: Ticknor & Fields.

Alcover, J. A., B. Seguí, and P. Bover. 1999. "Extinctions and Local Disappearances of Vertebrates in the Western Mediterranean Islands." Pp. 165–88 in *Extinctions in Near Time: Causes, Contexts, and Consequences*, edited by R. D. E. MacPhee. New York: Kluwer Academic/Plenum.

Allentoft, M. E., R. Heller, C. L. Oskam, E. D. Lorenzen, M. L. Hale, M. T. P. Gilbert, C. Jacomb, R. N. Holdaway, and M. Bunce. 2014. "Extinct New Zealand Megafauna Were Not in Decline Before Human Colonization." *Proceedings of the National Academy of Sciences* 111:4922–27. doi:10.1073/pnas.1314972111.

Alroy, J. 1999. "Putting North America's End-Pleistocene Megafaunal Extinction in Context: Large-Scale Analyses of Spatial Patterns, Extinction Rates, and Size Distributions." Pp. 105–43 in *Extinctions in Near Time: Causes, Contexts, and Consequences*, edited by R. D. E. MacPhee. New York: Kluwer Academic/Plenum.

———. 2001. "A Multispecies Overkill Simulation of the End-Pleistocene Megafaunal Mass Extinction." *Science* 292:1893–97.

Anderson, A., B. Allingham, and I. Smith, eds. 1996. *Shag River Mouth: The Archaeology of an Early Southern Maori Village. Research Papers in Archaeology and Natural History* 27. Canberra: Australian National University.

Anderson, A., T. H. Worthy, and R. McGovern-Wilson. 1996. "Moa Remains and Taphonomy." Pp. 200–213 in *Shag River Mouth: The Archaeology of an Early Southern Maori Village. Research Papers in Archaeology and Natural History* 27, edited by A. Anderson, B. Allingham, and I. Smith. Canberra: Australian National University.

Audubon, J. J. 1832. *Ornithological Biography, or an Account of the Habits of the Birds of the United States of America*. Philadelphia: Carey and Hart.

Bae, C. J., K. Douka, and M. D. Petraglia. 2017. "On the Origin of Modern Humans: Asian Perspectives." *Science* 358:eaai9067. doi:10.1126/science.aa9067.

Bahn, P., and J. Vertut. 1997. *Journey Through the Ice Age*. Berkeley: University of California Press.

Balouet, J.-C., and E. Alibert. 1990. *Extinct Species of the World*. New York: Barron's.

Barnosky, A. D., and E. L. Lindsey. 2010. "Timing of Quaternary Megafaunal Extinction in South America in Relation to Human Arrival and Climate Change." *Quaternary International* 217:10–29. doi:10.1016/j.quaint.2009.11.017.

Barnosky, A. D., N. Matzke, S. Tomiya, G. Wogan, B. Swartz, T. B. Quental, C. Marshall, J. L. McGuire, E. L. Lindsey, K. C. Maguire, B. Mersey, and E. A. Ferrer. 2011. "Has the Earth's Sixth Mass Extinction Already Arrived?" *Nature* 471:51–57. doi:10.1038/nature09678.

Bettinger, R. 1987. "Archaeological Approaches to Hunter-Gatherers." *Annual Review of Anthropology* 16:121–42.

Bobo, K. S., T. O. W. Kamgaing, E. C. Kamdoum, and Z. C. B. Dzefack. 2015. "Bushmeat Hunting in Southeastern Cameroon: Magnitude and Impact on Duikers (*Cephalophus* spp.)." *African Study Monographs*, Supplement 51:119–41. doi:10.14989/197202.

Borrero, L. A. 2009. "The Elusive Evidence: The Archaeological Record of the South American Extinct Megafauna." Pp. 145–68 in *American Megafaunal Extinctions at the End of the Pleistocene*, edited by G. Haynes. Dordrecht, Neth.: Springer Verlag.

Bourgeon L., A. Burke, and T. Higham. 2017. "Earliest Human Presence in North America Dated to the Last Glacial Maximum: New Radiocarbon Dates from Bluefish Caves, Canada." *PLoS ONE* 12 (1): e0169486. doi:10.1371/journal.pone.0169486.

Bradley, B., and D. Stanford. 2004. "The North Atlantic Ice-Edge Corridor: A Possible Palaeolithic Route to the New World." *World Archaeology* 36:459–78. doi:10.1080/0043824042000303656.

Braje, T. J., T. D. Dillehay, J. M. Erlandson, R. G. Klein, T. C. Rick. 2017. "Finding the First Americans." *Science* 358:592–94. doi:10.1126/science.aao5473.

Brook, B. W., and D. M. J. Bowman. 2002. "Explaining the Pleistocene Megafaunal Extinctions: Models, Chronologies, and Assumptions." *Proceedings of the National Academy of Sciences* 99:14624–27.

Brook, B. W., C. J. A. Bradshaw, A. Cooper, C. N. Johnson, T. H. Worthy, M. Bird, R. Gillespie, and R. G. Roberts. 2013. "Lack of Chronological Support for Stepwise Prehuman Extinctions of Australian Megafauna." *Proceedings of the National Academy of Sciences* 110:E3368. doi:10.1073/pnas.1309226110.

Buckland, W. 1824. "Reliquae diluviane"*: Observations on the Organic Remains Contained in Caves, Fissures, and Diluvial Gravel, and on Other Geological Phenomena, Attesting the Action of an Universal Deluge*, 2nd ed. London: John Murray.

———. 1831. "On the Occurrence of the Remains of Elephants, and Other Quadrupeds, in the Cliffs of Frozen Mud, in Eschscholtz Bay, Within Beering's Strait, and in Other Distant Parts of the Shores of the Arctic Seas." Pp. 593–612 in *Narrative of a Voyage to the Pacific and Beering's Strait to Co-operate with the Polar Expeditions, Performed in His Majesty's Ship* Blossom, *under the Command of Captain F. W. Beechey, R.N. . . . in the Years 1825, 26, 27, 28*, Part II, by F. W. Beechey. London: Henry Colburn and Richard Bentley.

Burney, D. A. 1993. "Late Holocene Environmental Changes in Arid Southwestern Madagascar." *Quaternary Research* 40:98–106.

———. 1999. "Rates, Patterns and Processes of Landscape Transformation and Extinction in Madagascar." Pp. 145–64 in *Extinctions in Near Time: Causes, Contexts, and Consequences*, edited by R. D. E. MacPhee. New York: Kluwer Academic/Plenum.

Burney, D. A., G. S. Robinson, and L. P. Burney. 2003. "*Sporormiella* and the Late Holocene Extinctions in Madagascar." *Proceedings of the National Academy of Sciences* 100:10800–10805. doi:10.1073/pnas.1534700100.

Campos, P. F., E. Willerslev, A. V. Sher, L. Orlando, E. Axelsson, A. N. Tikhonov, K. Aris-Sørensen, A. D. Greenwood, R.-D. Kahlke, P. Kosintsev, et al. 2010. "Ancient DNA Analyses Exclude Humans as the Driving Force Behind Late Pleistocene Musk Ox." *Proceedings of the National Academy of Science* 107:5675–80. doi:10.1073/pnas.0907189107.

Capriles, J. M., J. Albarracin-Jordan, U. Lombardo, D. Osorio, B. Maley, S. T. Goldstein, K. A. Herrera, M. D. Glascock, A. I. Domic, H. Veit, and C. M. Santoro. 2016. "High-Altitude Adaptation and Late Pleistocene Foraging in the Bolivian Andes." *Journal of Archaeological Science: Reports* 6:463–74. doi:10.1016/j.jasrep.2016.03.006.

Carleton, M. D., and S. L. Olson. 1999. "Amerigo Vespucci and the Rat of Fernando de Noronha: A New Genus and Species of Rodentia (Muridae, Sigmodontinae) from a Volcanic Island off Brazil's Continental Shelf." *American Museum Novitates* 3256:1–59.

Cartmill, M. 1993. *A View to a Death in the Morning: Hunting and Nature Through History*. Cambridge: Harvard University Press.

Chatters, J. C., D. J. Kennett, Y. Asmerom, B. M. Kemp, V. Polyak, A. Nava Blank, P. A. Beddows, E. Reinhardt, J. Arroyo-Cabrales, D. A. Bolnick, et al. 2014. "Late Pleistocene Human Skeleton and mtDNA Link Paleoamericans and Modern Native Americans." *Science* 344:750–54. doi:10.1126/science.1252619.

Clarkson, C., Z. Jacobs, B. Marwick, R. Fullagar, L. Wallis, M. Smith, R. G. Roberts, E. Hayes, K. Lowe, X. Carah, et al. 2017. "Human Occupation of Northern Australia by 65,000 Years Ago." *Nature* 547:306–10. doi:10.1038/nature22968.

Cooke, S., L. M. Dávalos, A. M. Mychajliw, S. T. Turvey, and N. S. Upham. 2017. "Anthropogenic Extinction Dominates Holocene Declines of West Indian Mammals." *Annual Review of Ecology, Evolution, and Systematics* 48:301–27. doi:10.1146/annrev-ecolsys-110316-022754.

Cooper, A., C. Turney, K. A. Hughen, B. W. Brook, H. G. McDonald, and C. J. A. Bradshaw. 2015. "Abrupt Warming Events Drove Late Pleistocene Holarctic Megafaunal Turnover." *Science* 349:602–6. doi:10.1126/science.aac4315.

Corfield, T. F. 1973. "Elephant Mortality in Tsavo National Park, Kenya." *East African Wildlife Journal* 11:339–68.

Cosgrove, R., J. Field, J. Garvey, J. Brenner-Coltrain, A. Goede, B. Charles, S. Wroe, A. Pike-Tay, R. Grun, M. Aubert, et al. 2010. "Overdone Overkill—The Archaeological Perspective on Tasmanian Megafaunal Extinctions." *Journal of Archaeological Science* 37:2486–503.

Cosgrove, R. and J. Garvey. 2017. "Behavioural Inferences from Late Pleistocene Aboriginal Australia: Seasonality, Butchery, and Nutrition in Southwest Tasmania." In *The Oxford Handbook of Zooarchaeology*, edited by U. Albarella, M. Rizzetto, H. Russ, K. Vickers, and S. Viner-Daniels. Oxford, Eng.: Oxford University Press. doi:10.1093/oxfordhb/9780199686476.013.49.

Croft, D. A. 2016. *Horned Armadillos and Rafting Monkeys: The Fascinating Fossil Mammals of South America*. Bloomington: University of Indiana Press.

Cuvier, G. 1829. *A Discourse on the Revolutions of the Surface of the Globe, and the Changes Thereby Produced in the Animal Kingdom*. London: Whittaker, Treacher, and Arnot.

Cuvier, G., and A. Brongniart. 1822. *Description géologique des Environs de Paris*. Paris: G. Dufour et E. D'Ocagne.

Darwin, C. R. 1839. *Journal of Researches into the Geology and Natural History of the Various Countries Visited by H.M.S.* Beagle*: Under the Command of Captain FitzRoy, R.N. from 1832 to 1836*. London: Henry Colburn.

DelGiudice, G. D. 1998. "Surplus Killing of White-Tailed Deer by Wolves in Northcentral Minnesota." *Journal of Mammalogy* 79:227–35.

Dewar, R. E., C. Radimilahy, H. T. Wright, Z. Jacobs, G. O. Kelly, and F. Bernag. 2013. "Stone Tools and Foraging in Northern Madagascar Challenge Holocene Extinction Models." *Proceedings of the National Academy of Sciences* 110:12583–88. doi:10.1073/pnas.1306100110.

Diamond, J. M. 1984. "Historical Extinctions: A Rosetta Stone for Understanding Prehistoric Extinctions." Pp. 824–62 in *Quaternary Extinctions: A Prehistoric Revolution*, edited by P. S. Martin and R. G. Klein. Tucson: University of Arizona Press.

Digby, B. 1926. *The Mammoth and Mammoth-Hunting in North-East Siberia*. London: H.F. & G. Witherby.

Dillehay, T. D., C. Ocampo, J. Saavedra, A. Oliviera Sawakuchi, R. M. Vega, M. Pino, M. B. Collins, L. C. Cummings, I. Arregui, X. S. Villagran, et al. 2015. "New Archaeological Evidence for an Early Human Presence at Monte Verde, Chile." *PLoS ONE* 10 (11): e0141923. doi:10.1371/journal.pone.0141923.

Dugatkin, L. A. 2009. *Mr. Jefferson and the Giant Moose: Natural History in Early America*. Chicago: University of Chicago Press.

Ehlers, J., P. D Hughes, and P. L. Gibbard. 2016. *The Ice Age*. New York: John Wiley & Sons.

Ember, C. R. 2014. "Hunter-Gatherers." In *Explaining Human Culture: Human Relations Area Files*, edited by C. R. Ember. http://hraf.yale.edu/ehc/summaries/hunter-gatherers.

Enk J., R. Debruyne, A. Devault, C. E. King, T. R. Terangen, D. O'Rourke, S. Salzburg, D. Fisher, R. D. E. MacPhee, and H. Poinar. 2011. "Complete Columbian Mammoth Mitogenome Suggests Interbreeding with Woolly Mammoths." *Genome Biology* 12:R51. doi:10.1186/gb-2011-12-5-r51.

Enk, J., A. Devault, C. Widga, J. Saunders, P. Szpak, J. Southon, J.-M. Rouillard, B. Shapiro, G. B. Golding, G. Zazula, et al. 2016. "*Mammuthus* Population Dynamics in Late Pleistocene North America: Divergence, Phylogeography, and Introgression." *Frontiers in Ecology and Evolution* doi:10.3389/fevo.2016.00042.

Faith, T. J., and T. A. Surovell. 2009. "Synchronous Extinction of North America's Pleistocene Mammals." *Proceedings of the National Academy of Sciences* 106:20641–45. doi:10.1073/pnas.0908153106.

Fariña, R. A., P. S. Tambusso, L. Varela, A. Czerwonogora, M. Di Giacomo, M. Musso, R. Bracco, and A. Gascue. 2013. "Arroyo del Vizcaino, Uruguay: A Fossil-Rich 30-ka-Old Megafaunal Locality with Cut-Marked Bones." *Proceedings of the Royal Society* B281:20132211. doi:10.1098/rspb.2013.2211.

Fariña, R. A., S. F. Vizcaíno, and G. De Iuliis. 2013. *Megafauna: Giant Beasts of Pleistocene South America*. Bloomington: University of Indiana Press.

Feranec, R. S., N. A. Miller, J. C. Lothrop, and R. Graham. 2011. "The *Sporormiella* Proxy and End-Pleistocene Megafaunal Extinction: A Perspective." *Quaternary International* 245:333–38. doi:10.1016/j.quaint.2011.06.004.

Fiedel, S. 2009. "Sudden Deaths: The Chronology of Terminal Pleistocene Megafaunal Extinctions." Pp. 21–38 in *American Megafaunal Extinctions at the End of the Pleistocene*, edited by G. Haynes. Dordrecht, Neth.: Springer Verlag.

Figuier, L. 1866. *The World Before the Deluge*. New York: D. Appleton.

Firestone R. B., A. West, J. P. Kennett, L. Becker, T. E. Bunch, Z. S. Revay, P. H. Schultz, T. Belgya, D. J. Kennett, J. M. Erlandson, et al. 2007. "Evidence for an Extraterrestrial Impact 12,900 Years Ago That Contributed to the Megafaunal Extinctions and the Younger Dryas Cooling." *Proceedings of the National Academy of Sciences* 104:16016–21. doi:10.1073/pnas.0706977104.

Flannery, T. 1995. *The Future Eaters*. New York: George Braziller.

Flannery, T., and P. Schouten. 2001. *A Gap in Nature: Discovering the World's Extinct Animals*. Melbourne, Aust.: Text Publishing.

Frison, G. C. 1998. "Paleoindian Large Mammal Hunters on the Plains of North America." *Proceedings of the National Academy of Sciences* 95:14576–83. doi:10.1073/pnas.95.24.14576.

Fuller, E. 2000. *Extinct Birds*. Oxford, Eng.: Oxford University Press.

Funder, S., O. Bennike, J. Bocher, C. Israelson, K. S. Petersen, and L. A. Simonarson. 2001. "Late Pliocene Greenland—The Kap Kobenhavn Formation in Greenland." *Bulletin of the Geological Society of Denmark* 48:117–34.

Godfrey, L. R., and W. L. Jungers. 2003. "The Extinct Sloth Lemurs of Madagascar." *Evolutionary Anthropology* 12:252–63.

Goodman, S. M., and W. L. Jungers. 2014. *Extinct Madagascar, Picturing the Island's Past*. Chicago: University of Chicago Press.

Gowdy, J. 1999. "Hunter-Gatherers and the Mythology of the Market." Pp. 391–98 in *The Cambridge Encyclopedia of Hunters and Gatherers*, edited by R. B. Lee and R. Daly. Cambridge, Eng.: Cambridge University Press.

Graham, R. W. 1985. "Response of Mammalian Communities to Environmental Changes During the Late Quaternary." Pp. 300–313 in *Community Ecology*, edited by J. Diamond and T. J. Chase. New York: Harper and Row.

Graham, R. W., and E. L. Lundelius. 1984. "Coevolutionary Disequilibrium and Pleistocene Extinctions."

Pp. 243–49 in *Quaternary Extinctions: A Prehistoric Revolution*, edited by P. S. Martin and R. G. Klein. Tucson: University of Arizona Press.

Gramling, C. 2016. "The 'Hobbit' Was a Separate Species of Human, New Dating Reveals." http://www.sciencemag.org/news/2016/03/hobbit-was-separate-species-human-new-dating-reveals. doi:10.1126/science.aaf9853.

Grayson, D. K. 1984. "Nineteenth-Century Explanations of Pleistocene Extinctions: A Review and Analysis." Pp. 5–39 in *Quaternary Extinctions: A Prehistoric Revolution*, edited by P. S. Martin and R. G. Klein. Tucson: University of Arizona Press.

Grayson, D. K., and D. J. Meltzer. 2003. "A Requiem for North American Overkill." *Journal of Archaeological Sciences* 30:585–93. doi:10.1016/S0305-4403(02)00205-4.

———. 2015. "Revisiting Paleoindian Exploitation of Extinct North American Mammals." *Journal of Archaeological Science* 56:177–93. doi:10.1016/j.jas.2015.02.009.

Guilday, J. E. 1967. "Differential Extinction During Late-Pleistocene and Recent Times." Pp. 12140 in *Pleistocene Extinctions: The Search for a Cause*, edited by P. S. Martin and H. E. Wright. New Haven: Yale University Press.

Guthrie, R. D. 1984. "Mosaics, Allelochemicals and Nutrients: An Ecological Theory of Late Pleistocene Megafaunal Extinctions." Pp. 259–98 in *Quaternary Extinctions: A Prehistoric Revolution*, edited by P. S. Martin and R. G. Klein. Tucson: University of Arizona Press.

———. 2005. *The Nature of Paleolithic Art*. Chicago: University of Chicago Press.

Haas, R., I. C. Stefanescu, A. Garcia-Putnam, M. S. Aldenderfer, M. T. Clementz, M. S. Murphy, C. Viviano Llave, and J. T. Watson 2017. "Humans Permanently Occupied the Andean Highlands by at Least 7 ka." *Royal Society Open Science* 4 (6): 170331. doi:10.1098/rsos.170331.

Hagstrum, J., R. B. Firestone, A. West, J. C. Weaver, and T. E. Bunch. 2017. "Impact-Related Microspherules in Late Pleistocene Alaskan and Yukon 'Muck' Deposits Signify Recurrent Episodes of Catastrophic Emplacement." *Scientific Reports* 7:16620. doi:10.1038/s41598-017-16958-2.

Haile, J., D. Froese, R. D. E. MacPhee, R. G. Roberts, L. J. Arnold, A. V. Reyes, M. Rasmussen, R. Nielsen, B. W. Brook, S. Robinson, et al. 2009. "Ancient DNA Reveals Late Survival of Mammoth and Horse in Interior Alaska." *Proceedings of the National Academy of Sciences* 106:22353–57. doi:10.1073/pnas.0912510106.

Halligan, J. J., M. R. Waters, and A. Perrotti. 2016. "Pre-Clovis Occupation 14,550 Years Ago at the Page-Ladson Site, Florida, and the Peopling of the Americas." *Science Advances* 2 (5): e1600375. doi:10.1126/sciadv.1600375.

Haynes, G. 2002. "The Catastrophic Extinction of North American Mammoths and Mastodonts," *World Archaeology* 33:391–416. doi:10.1080/00438240120107440.

———. 2009a. "Estimates of Clovis-Era Megafaunal Populations and Their Extinction Risks." Pp. 39–53 in *American Megafaunal Extinctions at the End of the Pleistocene*, edited by G. Haynes. Dordrecht, Neth.: Springer Verlag.

———, ed. 2009b. *American Megafaunal Extinctions at the End of the Pleistocene*. Dordrecht, Neth.: Springer Verlag.

Heintzman, P. D., D. Froese, J. W. Ives, A. E. R. Soares, G. D. Zazula, B. Letts, T. D. Andrews, J. C. Driver, E. Hall, P. G. Hare, et al. 2016. "Bison Phylogeography Constrains Dispersal and Viability of the Ice Free Corridor in Western Canada." *Proceedings of the National Academy of Sciences* 113:8057–63. doi:10.1073/pnas.1601077113.

Hiscock, O. 2008. *Archaeology of Ancient Australia*. New York: Routledge.

Holdaway, R. N., M. E. Allentoft, C. Jacomb, C. L. Oskam, N. R. Beavan, and M. Bunce. 2014. "An Extremely Low-Density Human Population Exterminated New Zealand Moa." *Nature Communications* 5:5436. doi:10.1038/ncomms6436.

Holder, C. F. 1886. *The Ivory King: A Popular History of the Elephant and Its Allies*. New York: Scribner's Sons.

Holen, S. R., T. A. Deméré, D. C. Fisher, R. Fullagar, J. B. Paces, G. T. Jefferson, J. M. Beeton, R. A. Cerutti, A. N. Rountrey, L. Vescera, and K. A. Holen. 2017. "A 130,000-Year-Old Archaeological Site in Southern California, USA." *Nature* 544:479–83. doi:10.1038/nature22065.

Holliday, V. T. 2015. "Problematic Dating of Claimed Younger Dryas Boundary Impact Proxies." *Proceedings of the National Academy of Sciences*. doi:10.1073/pnas.1518945112.

Horan, R. D., J. F. Shogren, and E. Bulte. 2003. "A Paleoeconomic Theory of Co-evolution and Extinction of Domesticable Animals." *Scottish Journal of Political Economy* 50:131–48. doi:10.1111/1467-9485.5002002.

Hublin, J-J., A. Ben-Ncer, S. E. Bailey, S. E. Freidline, S. Neubauer, M. M. Skinner, I. Bergmann, A. Le Cabec, S. Benazzi, K. Harvati, and P. Gunz. 2017. "New Fossils from Jebel Irhoud, Morocco and the Pan-African Origin of *Homo sapiens*." *Nature* 546:289–92. doi:10.1038/nature22336.

Humbert, H. 1927. "La destruction d'une flore insulaire par le feu: Principaux aspects de la végétation à Madagascar." *Mémoires de l'Academie Malgache* 5:1–80.

Hyslop, J., ed. 1988. *Travels and Archaeology in South Chile by Junius B Bird, with Journal Segments by Margaret Bird*. Iowa City: Iowa University Press.

Ingold, T. I. 1999. "On the Social Relations of the Hunter-Gatherer Band." Pp. 399–410 in *The Cambridge Encyclopedia of Hunters and Gatherers*, edited by R. B. Lee and R. Daly. Cambridge, Eng.: Cambridge University Press.

Jablonski, D. 2001. "Lessons from the Past: Evolutionary Impacts of Mass Extinctions." *Proceedings of the National Academy of Sciences* 98:5393–98. doi:10.1073ypnas.101092598.

Johnson, C. N. 2002. "Determinants of Loss of Mammal Species During the Late Quaternary 'Megafaunal' Extinctions: Life History and Ecology, but Not Body Size." *Proceedings of the Royal Society of London* B269:2221–22.

———. 2006. *Australia's Mammal Extinctions: A 50,000 Year History*. Cambridge, Eng.: Cambridge University Press.

———. 2009. "Ecological Consequences of Late Quaternary Extinctions of Megafauna." *Proceedings of the Royal Society of London* B276:2509–19. doi:10.1098/rspb.2008.1921.

Kelly, R. L. and M. M. Prasciunas. 2007. "Did the Ancestors of Native Americans Cause Animal Extinctions in Late-Pleistocene North America? And Does It Matter If They Did?" Pp. 95–122 in *Native Americans and the Environment: Perspectives on the Ecological Indian*, edited by M. E. Harkin and D. R. Lewis. Lincoln: University of Nebraska Press.

Kennett, J. P., D. J. Kennett, B. J. Culleton, J. E. Aura Tortosa, J. L. Bischoff, T. E. Bunch, I. R. Daniel, J. M. Erlandson, D. Ferraro, R. B. Firestone, et al. 2015. "Bayesian Chronological Analyses Consistent with Synchronous Age of 12,835–12,735 cal B.P. for Younger Dryas Boundary on Four Continents." *Proceedings of the National Academy of Sciences* 112:E4344–E4353. doi:10.1073/pnas.1507146112.

Kock, R. A., M. Orynbayev, S. Robinson, S. Zuther, N. J. Singh, W. Beauvais, E. R. Morgan, A. Kerimbayev, S. Khomenko, H. M. Martineau, et al. 2018. "Saigas on the Brink: Multidisciplinary Analysis of the Factors Influencing Mass Mortality Events." *Science Advances* 4 (1): eaao2314. doi:10.1126/sciadv.aao2314.

Kolbert, E. 2014. *The Sixth Extinction: An Unnatural History*. New York: Henry Holt.

Kruuk, H. 1972. "Surplus Killing by Carnivores." *Journal of Zoology* 166:233–44.

Lee, R. B., and R. Daly, eds. 1999. *The Cambridge Encyclopedia of Hunters and Gatherers*. Cambridge, Eng.: Cambridge University Press.

Leydet, D. J., A. E. Carlson, J. T. Teller, A. Breckenridge, A. M. Barth, D. J. Ullman, G. Sinclair, G. A. Milne, J. K. Cuzzone, and M. W. Caffee. 2018. "Opening of Glacial Lake Agassiz's Eastern Outlets by the Start of the Younger Dryas Cold Period." *Geology* 46:155–58. doi:10.1130/G39501.

Lister, A., and P. Bahn. 2015. *Mammoths: Giants of the Ice Age*, rev. ed. Edison, N.J.: Chartwell Books.

Llamas, B., L. Fehren-Schmitz, G. Valverde, J. Soubrier, S. Mallick, N. Rohland, S. Nordenfelt, C. Valdiosera, S. M. Richards, A. Rohrlach, et al. 2016. "Ancient Mitochondrial DNA Provides High-Resolution Time Scale of the Peopling of the Americas." *Science Advances* 2 (4): e1501385. doi:10.1126/sciadv.1501385.

Long, J. A., M. Archer, T. Flannery, and S. Hand. 2003. *Prehistoric Mammals of Australia and New Guinea: One Hundred Million Years of Evolution*. Baltimore: Johns Hopkins University Press.

Louys, J., D. Curnoe, and H. Tong. 2007. "Characteristics of Pleistocene Megafaunal Extinctions in Southeast Asia." *Palaeogeography, Palaeoclimatology, Palaeoecology* 243:152–73.

Lyell, C. 1866. *Geological Evidences for the Antiquity of Man*. London: John Murray.

Lyons, K., F. A. Smith, and J. H. Brown. 2004a. "Of Mice, Mastodons, and Men: Human Mediated Extinctions on Four Continents." *Evolutionary Ecology Research* 6:339–58.

Lyons, K., F. A. Smith, P. J. Wagner, E. P. White, and J. H. Brown. 2004b. "Of Mice, Mastodons, and Men: Human Mediated Extinctions on Four Continents." *Evolutionary Letters* 7:859–68. doi:10.1111/j.1461-0248.2004.00643.x.

MacDonald, F. 2016. "Study Proves the Explorers Club Didn't Really Eat Mammoth at 1950s New York Dinner." http://www.sciencealert.com/study-proves-the-explorers-club-didn-t-really-eat-mammoth-at-1950s-new-york-dinner.

MacPhee, R. D. E., ed. 1999. *Extinctions in Near Time: Causes, Contexts, and Consequences*. New York: Kluwer Academic/Plenum.

———. 2009. "*Insulae infortunatae*: Establishing a Chronology for Late Quaternary Mammal Extinctions in the West Indies." Pp. 169–94 in *American Megafaunal Extinctions at the End of the Pleistocene*, edited by G. Haynes. Dordrecht, Neth.: Springer Verlag.

MacPhee, R. D. E., and D. A. Burney. 1991. "Dating of Modified Femora of Extinct Dwarf *Hippopotamus* from Southern Madagascar: Implications for Constraining Human Colonization and Vertebrate Extinction Events." *Journal of Archaeological Science* 18:695–706.

MacPhee, R. D. E., and C. Flemming. 1999. "*Requiem aeternum*: The Last Five Hundred Years of Mammalian Species Extinctions." Pp. 333–71 in *Extinctions in Near Time: Causes, Contexts, and Consequences*, edited by R. D. E. MacPhee. New York: Kluwer Academic/Plenum.

MacPhee, R. D. E., and A. D. Greenwood. 2013 "Infectious Disease, Endangerment, and Extinction." *International Journal of Evolutionary Biology* 2013. doi:10.1155/2013/571939

MacPhee, R. D. E., M. A. Iturralde-Vinent, and O. Jiménez-Vázquez. 2007. "Prehistoric Sloth Extinctions in Cuba: Implications of a New 'Last' Appearance Date." *Caribbean Journal of Science* 41:94–98.

MacPhee, R. D. E., and P. A. Marx. 1997. "The 40,000-Year Plague: Humans, Hyperdisease, and First-Contact Extinctions." Pp. 169–217 in *Natural Change and Human Impact in Madagascar*, edited by S. Goodman and B. Patterson. Washington, D.C.: Smithsonian Institution Press.

MacPhee, R. D. E., A. N. Tikhonov, D. Mol, C. de Marliave, H. van der Plicht, A. D. Greenwood, C. Flemming, and L. Agenbroad. 2002. "Radiocarbon Chronologies and Extinction Dynamics of the Late Quaternary Mammalian Megafauna of the Taimyr Peninsula, Russian Federation." *Journal of Archaeological Science* 29:1017–42.

Mahaney, W. C., D. H. Krinsley, V. Kalm, K. Langworthy, and J. Ditto. 2011. "Notes on the Black Mat Sediment, Mucuñuque Catchment, Northern Mérida Andes, Venezuela." *Journal of Advanced Microscopy Research* 6:1–9.

Mao, F. 2017. "Tasmanian Tigers Were in Poor Genetic Health, Study Finds." http://www.bbc.com/news/world-australia-42318444.

Marlow, C. 2000. *The Ghosts of Evolution: Nonsensical Fruit, Missing Partners, and Other Ecological Anachronisms*. New York: Basic Books.

Marshall, C. R., E. L. Lindsey, N. A. Villavicencio, and A. D. Barnosky. 2015. "A Quantitative Model for Distinguishing Between Climate Change, Human Impact, and Their Synergistic Interaction as Drivers of the Late Quaternary Megafaunal Extinctions." Pp. 1–20 in *Earth-Life Transitions: Paleo biology in the Context of Earth System Evolution. The Paleontological Society Papers* 21, edited by D. Polly J. J. Head, and D. L. Fox.

Martin, P. S. 1967. "Prehistoric Overkill." Pp. 75–120 in *Pleistocene Extinctions: The Search for a Cause*, edited by P. S. Martin and H. E. Wright. New Haven: Yale University Press.

———. 1973. "The Discovery of America." *Science* 179:969–74.

———. 1984. "Prehistoric Overkill: The Global Model." Pp. 354–403 in *Quaternary Extinctions: A Prehistoric Revolution*, edited by P. S. Martin and R. G. Klein, Tucson: University of Arizona Press.

———. 1990. "40,000 Years of Extinction on the 'Planet of Doom.'" *Palaeogeography, Palaeoclimatology, Palaeoecology* 82:187–201.

———. 2005. *Twilight of the Mammoths: Ice Age Extinctions and the Rewilding of America*. Berkeley: University of California Press.

Martin, P. S., and R. G. Klein, eds. 1984. *Quaternary Extinctions: A Prehistoric Revolution*. Tucson: University of Arizona Press.

Martin, P. S., and D. W. Steadman. 1999. "Prehistoric Extinctions on Islands and Continents." Pp. 17–55 in *Extinctions in Near Time: Causes, Contexts, and Consequences*, edited by R. D. E. MacPhee. New York: Kluwer Academic/Plenum.

Martin, P. S., and H. E. Wright, eds. 1967. *Pleistocene Extinctions: The Search for a Cause*. New Haven: Yale University Press.

Mayor, A. 2000. *The First Fossil Hunters: Paleontology in Greek and Roman Times*. Princeton, N.J.: Princeton University Press.

———. 2005. *Fossil Legends of the First Americans*. Princeton, N.J.: Princeton University Press.

McGlone, M. 2012. "The Hunters Did It." *Science* 335:1452–53. doi:10.1126/science.1220176.

Meltzer, D. J. 2010. *First People in a New World: Colonizing Ice Age America*. Berkeley: University of California Press.

———. 2015. "Pleistocene Overkill and North American Mammalian Extinctions." *Annual Review of Anthropology* 44:33–53. doi:10.1146/annurev-anthro-102214-013854.

Metcalf, J. L., C. Turney, R. Barnett, F. Martin, S. C. Bray, J. T. Vilstrup, L. Orlando, R. Salas-Gismondi, D. Loponte, M. Medina, et al. 2016. "Synergistic Roles of Climate Warming and Human Occupation in Patagonian Megafaunal Extinctions During the Last Deglaciation." *Science Advances* 2 (6) :e1501682. doi:10.1126/sciadv.1501682.

Millener, P. R. 1988 "Contributions to New Zealand's Late Quaternary Avifauna. 1: *Pachyplichas*, a New Genus of Wren (Aves: Acanthisittidae), with Two New Species," *Journal of the Royal Society of New Zealand* 18:383–406. doi:10.1080/03036758.1988.10426464.

Mlot, C. 2017. "Two Wolves Survive in World's Longest Running Predator-Prey Study." *Science*. doi:10.1126/science.aal1061.

Montulé, E. 1821. *A Voyage to North America and the West Indies in 1817*. London: Richard Phillips.

Mosimann, J. E., and P. S. Martin. 1975. "Simulating Overkill by Paleoindians." *American Scientist* 63:304–13.

Murray, G. G. R., A. E. R. Soares, B. J. Novak, N. K. Schaefer, J. A. Cahill, A. J. Baker, J. R. Demboski, A. Doll, R. R. Da Fonseca, T. L. Fulton, et al. 2017. "Natural Selection Shaped the Rise and Fall of Passenger Pigeon Genomic Diversity." *Science* 358:951–54. doi:10.1126/science.aao0960.

Naish, D. 2009. "The Small, Recently Extinct, Island-Dwelling Crocodilians of the South Pacific." http://scienceblogs.com/tetrapodzoology/2009/05/13/mekosuchines-2009/.

Naito, Y. I., M. Germonpré, Y. Chikaraishi, N. Ohkouchi, D. G. Drucker, K. A. Hobson, M. A. Edwards, C. Wissing, and H. Bocherens. 2016. "Evidence for Herbivorous Cave Bears (*Ursus spelaeus*) in Goyet Cave, Belgium: Implications for Palaeodietary Reconstruction of Fossil Bears Using Amino Acid $\delta^{15}N$ Approaches." *Journal of Quaternary Science* 31:598–606. doi:10.1002/jqs.2883.

Nicholls, H. 2015. "Mysterious Die-Off Sparks Race to Save *Saiga* Antelope." *Nature*. doi:10.1038/nature.2015.17675.

Nogués-Bravo, D., J. Rodríguez, J. Hortal, P. Batra, and M. B. Araújo. 2008. "Climate Change, Humans, and the Extinction of the Woolly Mammoth." *PLoS Biology* 6 (4): e79. doi:10.1371/journal.pbio.0060079.

Nowak, R. M. 1999. *Walker's Mammals of the World*, 6th ed., 2 vols. Baltimore: Johns Hopkins University Press.

Oppo, D. W., and W. B. Curry. 2012. "Deep Atlantic Circulation During the Last Glacial Maximum and Deglaciation." *Nature Education Knowledge* 3 (10). https://www.nature.com/scitable/knowledge/library/deep-atlantic-circulation-during-the-last-glacial-25858002.

Osborn, H. F. 1906. "The Causes of Extinction of Mammalia." *American Naturalist* 40:769–95, 829–59.

———. 1910. *The Age of Mammals in Europe, Asia, and North America*. New York: Macmillan.

Owen, R. 1844. "On *Dinornis*, an Extinct Genus of Tridactyle Struthious Birds, with Descriptions of Portions of the Skeleton of Five Species Which Formerly Existed in New Zealand." *Transactions of the Zoological Society of London* 3:235–75.

Owen-Smith, N. 1999. "The Interaction of Humans, Megaherbivores and Habitats in the Late Pleistocene Extinction Event." Pp. 57–69 in *Extinctions in Near Time: Causes, Contexts, and Consequences*, edited by R. D. E. MacPhee. New York: Kluwer Academic/Plenum.

Panagopoulou, E., V. Tourloukis, N. Thompson, A. Athanassiou, G. Tsartsidou, G. E. Konidaris, D. Giusti, P. Karkanas, and K. Harvati. 2015. "Marathousa 1: A New Middle Pleistocene Archaeological Site from Greece." *Antiquity* 89 (343). https://www.antiquity.ac.uk/projgall/panagopoulou343.

Pedersen, M. W., A. Ruter, C. Schweger, H. Friebe, R. A. Staff, K. K. Kjeldsen, M. L. Z. Mendoza, A. B. Beaudoin, C. Zutter, N. K. Larsen, et al. 2016. "Postglacial Viability and Colonization in North America's Ice-Free corridor. *Nature* 537:45–49. doi:10.1038/nature19085.

Pitulko, V. V., A. N. Tikhonov, E. Y. Pavlova, P. A. Nikolskiy, K. E. Kuper, and R. N. Polozov. 2016. "Early Human Presence in the Arctic: Evidence from 45,000-Year-Old Mammoth remains." *Science* 351:260–63. doi:10.1126/science.aad0554.

Politis, G., M. A. Gutiérrez, D. J. Rafuse, and A. Blasi. 2016. "The Arrival of *Homo sapiens* into the Southern Cone at 14,000 Years Ago." *PLoS ONE* 11 (9): e0162870. doi:10.1371/journal.pone.0162870.

Powell, W. 2016. "New Genetically Engineered American Chestnut Will Help Restore the Decimated, Iconic Tree." *The Conversation*. http://theconversation.com/new-genetically-engineered-american-chestnut-will-help-restore-the-decimated-iconic-tree-52191.

Prideaux, G., J. A. Long, L. K. Ayliffe, J. C. Hellstrom, B. Pillans, W. E. Boles, M. N. Hutchinson, R. G. Roberts, M. L. Cupper, L. J. Arnold, et al. 2007. "An Arid-Adapted Middle Pleistocene Vertebrate Fauna from South-Central Australia." *Nature* 445:422–25. doi:10.1038/nature05471.

Pushkina, D., H. Bocherens, and R. Ziegler. 2014. "Unexpected Palaeoecological Features of the Middle and Late Pleistocene Large Herbivores in Southwestern Germany Revealed by Stable Isotopic Abundances in Tooth Enamel." *Quaternary International* 339/340:164–78. doi:10.1016/j.quaint.2013.12.033.

Quammen, D. 1996. *The Song of the Dodo: Island Biogeography in an Age of Extinctions*. New York: Scribner.

———, ed. 2008. *Charles Darwin: On the Origin of Species, the Illustrated Edition*. New York: Sterling.

Raia, P., C. Barbera, and M. Conte. 2003. "The Fast Life of a Dwarfed Giant." *Evolutionary Ecology* 17:293–312.

Roberts, P., C. Hunt, M. Arroyo-Kalin, D. Evans, and N. Boivin. 2017. "The Deep Human Prehistory of Global Tropical Forests and Its Relevance for Modern Conservation." *Nature Plants* 3 (8): 17093. doi: 10.1038/nplants.2017.93.

Rodgers, R., and M. Slatkin. 2016. "Excess of Genomic Defects in a Woolly Mammoth on Wrangel Island." *PLoS Genetics* 13 (3): e1006601. doi:10/1371/journal.pgen.1006601.

Romer, A. S. 1933. "Pleistocene Vertebrates and Their Bearing on the Problem of Human Antiquity in North America." Pp. 49–83 in *The American Aborigines, Their Origin and Antiquity*, edited by D. Jenness. Toronto: University of Toronto Press.

Rudwick, M. 1976. *The Meaning of Fossils: Episodes in the History of Palaeontology*, 2nd ed. Chicago: University of Chicago Press.

Rule, S., B. W. Brook, S. G. Haberle, C. S. M. Turney, A. P. Kershaw, and C. N. Johnson. 2012. "The Aftermath of Megafaunal Extinction: Ecosystem Transformation in Pleistocene Australia." *Science* 335:1483–86. doi:10.1126/science.1214261.

Russell, S. A. 1995. "The Pleistocene Extinctions: A Bedtime Story." *The Missouri Review* 18:30–39.

Saltré, F., C. Johnson, and C. Bradshaw. 2016. "Climate Change Not to Blame for Late Quaternary Megafauna Extinctions in Australia." *Nature Communications* 7:10511. doi:10.1038/ncomms10511.

Sánchez-Villagra, M., M. Geiger, and R. A. Schneider. 2016. "The Taming of the Neural Crest: A Developmental Perspective on the Origins of Morphological Co-variation in Domesticated Animals." *Royal Society Open Science* 3:160107. doi:10.1098/rsos.160107.

Sandom, C., S. Faurby, B. Sandel, and J.-C. Svenning. 2014. "Global Late Quaternary Megafauna Extinctions Linked to Humans, Not Climate Change." *Proceedings of the Royal Society* B281:20133254. doi:10.1098/rspb.2013.3254.

Segura, A. M., R. A. Fariña, and M. Arim. 2016. "Exceptional Body Sizes but Typical Trophic Structure in a Pleistocene Food Web." *Biology Letters* 12:20160228. doi:10.1098/rsbl.2016.0228.

Semonin, P. 2000. *American Monster: How the Nation's First Prehistoric Creature Became a Symbol of National Identity*. New York: New York University Press.

Shapiro, B. 2016. *How to Clone a Mammoth: The Science of De-Extinction*. Princeton, N.J.: Princeton University Press.

Short, J., J. E. Kinnear, and A. Robley. 2002. "Surplus Killing by Introduced Predators in Australia—Evidence for Ineffective Anti-Predator Adaptations in Native Prey Species?" *Biological Conservation* 103:283–301. pii:S0006-3207(01)00139-2.

Simmons, A. H. 1999. *Faunal Extinction in an Island Society*. New York: Kluwer Academic/Plenum.

Slavenko, A., O. J. S. Tallowin, Y. Itescu, P. Raia, and S. Mieri. 2016. "Late Quaternary Reptile Extinction: Size Matters, Insularity Dominates." *Global Ecology and Biogeography* 25:1308–20. doi:10.1111/geb.12491.

Smith, F., R. E. S. K. Lyons, and J. L Payne. 2018. "Body Size Downgrading of Mammals over the Late Quaternary." *Science* 360:310–13. doi:10.1126/science.aao5987.

Starkovich, B. M., and N. J. Conrad. 2015. "Bone Taphonomy of the Schöningen 'Spear Horizon South' and Its Implications for Site Formation and Hominin Meat Provisioning." *Journal of Human Evolution* 89:154–71. doi:10.1016/j.jhevol.2015.09.015.

Steadman, D. W. 2006. *Extinction and Biogeography of Tropical Pacific Birds*. Chicago: University of Chicago Press.

Steadman, D. W., P. S. Martin, R. D. E. MacPhee, A. J. T. Jull, H. G. McDonald, C. A. Woods, M. A. Iturralde-Vinent, and G. Hodgkins. 2005. "Asynchronous Extinction of Late Quaternary Sloths on Continents and Islands." *Proceedings of the National Academy of Sciences* 102:11763–68. doi:10.1073/pnas.0502777102.

Steadman, D. W., J. P. White, and J. Allen. 1999. "Prehistoric Birds from New Ireland, Papua New Guinea:

Extinctions on a Large Melanesian Island." *Proceedings of the National Academy of Sciences* 96:2563–68. doi:10.1073/pnas.96.5.2563.

Stuart, A. J. 1999. "Late Pleistocene Megafaunal Extinctions: A European Perspective." Pp. 257–69 in *Extinctions in Near Time: Causes, Contexts, and Consequences*, edited by R. D. E. MacPhee. New York: Kluwer Academic/Plenum.

Stuart, C. T. 1986. "The Incidence of Surplus Killing by *Panthera pardus* and *Felis caracal* in Cape Province, South Africa." *Mammalia* 50:556–58.

Tennyson, A., and P. Martinson. 2006. *Extinct Birds of New Zealand*. Wellington, N.Z.: Te Papa Press.

Thomas, M. A., G. W. Roemer, C. J. Donlan, B. G. Dickson, M. Matocq, and J. Malaney. 2013. "Ecology: Gene Tweaking for Conservation." *Nature* 501:485–86. doi:10.1038/501485a.

Tilesius, W. G. 1815. "De skeleto mammonteo Sibirico ad maris glacialis littora anno 1897 [sic], effosso, cui praemissae Elephantini generis specierum distinctiones." *Mémoires de l'Académie Impériale des Sciences de St. Pétersbourg* 5:406–514.

Turvey, S. T. 2009a. "In the Shadow of the Megafauna: Prehistoric Mammal and Bird Extinctions Across the Holocene." Pp. 17-39 in *Holocene Extinctions*, edited by S. T. Turvey. Oxford, Eng.: Oxford University Press.

———, ed. 2009b. *Holocene Extinctions*. Oxford, Eng.: Oxford University Press.

Turvey, S. T., H. Tong, A. J. Stuart, and A. M. Lister. 2013. "Holocene Survival of Late Pleistocene Megafauna in China: A Critical Review of the Evidence." *Quaternary Science Reviews* 76:156–66. doi:10.1016/j.quascirev.2013.06.030.

Vågene, Å. J., A. Herbig, M. G. Campana, N. M. Robles García, C. Warinner, S. Sabin, M. A. Spyrou, A. A. Valtueña, D. Huson, N. Tuross, et al. 2018. "*Salmonella enterica* Genomes from Victims of a Major Sixteenth-Century Epidemic in Mexico." *Nature Ecology and Evolution*. doi:10.1038/s41559-017-0446-6.

Van den Bergh, G. D., B. Mubroto, F. Aziz, P. Sondaar, and J. de Vos. 1996a. "Did *Homo erectus* Reach the Island of Flores?" *Bulletin of the Indo-Pacific Prehistory Association* 14:27–36.

Van den Bergh, G. D., P. Sondaar, J. de Vos, and F. Aziz. 1996b. "The Proboscideans of the South-East Asian Islands." Pp. 240–48 in *The Proboscidea: Evolution, Palaeoecology of Elephants and Their Relatives*, edited by J. Shoshoni and P. Tassy. Oxford, Eng.: Oxford University Press.

Van der Geer, A. A. E., G. A. Lyras, J. De Vos, and M. Dermitzakis. 2010. *Evolution of Island Mammals: Adaptation and Extinction of Placental Mammals on Islands*. New York: John Wiley & Sons.

Van der Geer, A. A. E., G. A. Lyras, L. W. van den Hoek Ostende, J. de Vos, and H. Drinia. 2014. "A Dwarf Elephant and a Rock Mouse on Naxos (Cyclades, Greece) with a Revision of the Palaeozoogeography of the Cycladic Islands During the Pleistocene." *Palaeogeography, Palaeoclimatology, Palaeoecology* 404:133–44.

Van der Geer, A. A. E., G. D. van den Bergh, G. A. Lyras, U. W. Prasetyo, R. Due Awe, E. Setiyabudi, and H. Drinia. 2016. "The Effect of Area and Isolation on Insular Dwarf Proboscideans." *Journal of Biogeography* 43:1656–66.

Vartanyan, S. L., V. E. Garutt, and A. V. Sher. 1993. "Holocene Dwarf Mammoths from Wrangel Island in the Siberian Arctic." *Nature* 362:337–49. doi:10.1038/362337a0.

Veltre, D. W., D. R. Yesner, K. J. Crossen, R. W. Graham, and J. B. Coltrain. 2008. "Patterns of Faunal Extinction and Paleoclimatic Change from Mid-Holocene Mammoth and Polar Bear Remains, Pribilof Islands, Alaska." *Quaternary Research* 70:40-50. doi:10.1016/j.yqres.2008.03.006.

Virah-Sawmy, M., K. J. Willis, and L. Gillson. 2010. "Evidence for Drought and Forest Declines During the Recent Megafaunal Extinctions in Madagascar." *Journal of Biogeography* 37 (3): 506–19. doi:10.1111/j.1365-2699.2009.02203.x.

Waguespack, N., and T. A. Surovell. 2003. "How Many Elephant Kills Are 14? Clovis Mammoth and Mastodon Kills in Context." *Quaternary International* 191:82–97. doi:10.1016/j.quaint.2007.12.001.

Waldren, A., and E. L. Layard. 1872. "On Birds Recently Observed or Obtained in the Island of Negros, Philippines." *The Ibis*, 3rd ser., 2:93–107.

Wallace, A. R. 1876. *The Geographical Distribution of Animals, with a Study of the Relations of Living and Extinct Faunas as Elucidating the Past Changes of the Earth's Surface.* New York: Harper & Brothers.

Waterhouse, G. R. 1839. "Part II. Mammalia, with a Notice of Their Habits and Ranges by C. Darwin." Pp. 1–97 in *The Zoology of the Voyage of H.M.S.* Beagle*: Under the Command of Captain FitzRoy, R.N. from 1832 to 1836*, edited by C. R. Darwin. London: Smith, Elder & Co.

Waters, M. R., T. W. Stafford, B. Kooyman, and L. V. Hills. 2015. "Late Pleistocene Horse and Camel Hunting at the Southern Margin of the Ice-Free Corridor: Reassessing the Age of Wally's Beach, Canada." *Proceedings of the National Academy of Sciences* 114:4263–67. doi:10.1073/pnas.1420650112.

Westaway, K. E., J. Louys, R. Due Awe, M. J. Morwood, G. J. Price, J.-X. Zhao, M. Aubert, R. Joannes-Boyau, T. M. Smith, M. M. Skinner, et al. 2017. "An Early Modern Human Presence in Sumatra 73,000–63,000 Years Ago." *Nature* 548:322–25. doi:10.1038/nature23452.

White, A. W., T. H. Worthy, S. Hawkins, S. Bedford, and M. Spriggs. 2010. "Megafaunal Meiolaniid Horned Turtles Survived until Early Human Settlement in Vanuatu, Southwest Pacific." *Proceedings of the National Academy of Sciences* 107:15512–16. doi:10.1073/pnas.1005780107.

Whitney-Smith, E. 2009. *The Second-Order Predation Hypothesis of Pleistocene Extinctions: A System Dynamics Model.* Saarbrücken, Ger.: VDM Verlag Dr Müller.

Wilson, A. M., T. Y. Hubel, S. D. Wilshin, J. C. Lowe, M. Lorenc, O. P. Dewhirst, H. L. A. Bartlam-Brooks, R. Diack, E. Bennitt, K. A. Golabek, et al. 2018. "Biomechanics of Predator-Prey Arms Race in Lion, Zebra, Cheetah and Impala." *Nature* 554:183–88. doi:10.1038/nature25479.

Worthy, T. H., and R. N. Holdaway. 2002. *The Lost World of the Moa.* Bloomington: Indiana University Press.

Wroe, S., J. Field, R. Fullagar, and L. S. Jermin. 2004. "Megafaunal Extinction in the Late Quaternary and the Global Overkill Hypothesis." *Alcheringa* 28:291–331.

Wyatt, K. B., P. Campos, M. T. P. Gilbert, W. H. Hynes, R. DeSalle, P. Daszak, S. Ball, R. D. E. MacPhee, and A. D. Greenwood. 2008. "Historical Mammal Extinction Due to Introduced Infectious Disease." *PLoS One* 3 (11): e3602.

Zazula, G. D., R. D. E. MacPhee, J. Z. Metcalfe, A. V. Reyes, F. Brock, P. S. Drukenmiller, P. Groves, C. R. Harington, G. W. L. Hodgins, M. L. Kunz, et al. 2014. "American Mastodon Extirpation in the Arctic and Subarctic Predates Human Colonization and Terminal Pleistocene Climate Change." *Proceedings of the National Academy of Sciences* 111:18460–65. doi:10.1073/pnas.1416072111.

Zazula, G. D., R. D. E. MacPhee, J. R. Southon, S. Nalawade-Chavan, and E. Hall. 2017. "A Case of Early Wisconsinan 'Over-chill': New Radiocarbon Evidence for Early Extirpation of Western Camel (*Camelops hesternus*) in Eastern Beringia." *Quaternary Science Reviews* 171:48–57.

Zimov, S. A. 2005. "Pleistocene Park: Return of the Mammoth's Ecosystem." *Science* 308:796–98. doi:10.1126/science.1113442.

Zimov, S. A., V. I. Chuprynin, A. P. Oreshko, F. S. Chapin, J. F. Reynolds, and M. C. Chapin. 1995. "Steppe-Tundra Transition: A Herbivore-Driven Biome Shift at the End of the Pleistocene." *American Naturalist* 146:765–94.

Zutovski, K., and R. Barkai. 2016. "The Use of Elephant Bones for Making Acheulian Handaxes: A Fresh Look at Old Bones." *Quaternary International* 406:227–38. doi:10.1016/quaint.2015.01.033.

GUIDE TO ADDITIONAL READING

Treatment of most of the topics connected with Near Time extinctions can be found in abundance on the internet. (An especially good resource is the Wikipedia article "Quaternary extinction event," https://en.wikipedia.org/wiki/Quaternary_extinction_event, to which I am not a contributor.) For those inclined to dig deeper, the contest of ideas over the cause of these losses continues to be vigorously conducted in leading general scientific journals such as *Proceedings of the National Academy of Sciences*, *Science*, and *Nature*, as well as a host of more specialized outlets too numerous to mention by title.

A number of well-illustrated volumes provide broad overviews of Near Time megafauna in various parts of the world. Among those published in the last decade and a half are *Australia's Mammal Extinctions* (2006) by Chris Johnson; *Megafauna: Giant Beasts of Pleistocene South America* (2013) by Richard Fariña, Sergio Vizcaíno, and Gerry De Iuliis; *Extinct Madagascar* (2014) by Steve Goodman and Bill Jungers; *Mammoths: Giants of the Ice Age* (2015) by Adrian Lister and Paul Bahn; and *Horned Armadillos and Rafting Monkeys* (2016) by Darin Croft. An excellent introduction to Quaternary climate and geology is *The Ice Age* (2016) by Jürgen Ehlers, Philip D. Hughes, and Philip L. Gibbard. If you enjoy Peter Schouten's art as much as I do, see any of the well illustrated volumes that he and co-authors have published in recent years. My favorite is *A Gap in Nature: Discovering the World's Extinct Animals* (2001), produced in collaboration with Tim Flannery. Complete publication data on these references may be found in References Cited.

ACKNOWLEDGMENTS

I cannot find sufficient words to thank my wife, Clare Flemming, who did all things necessary to keep our life on the rails, and much else besides, while I selfishly spent most of my free moments on the present project. I next thank Peter Schouten, whose artwork graces these pages, for his willingness to keep this project going despite our early setbacks. My agent, Gillian MacKenzie, and my fellow biologist and accomplished writer Bill Schutt together expressed sufficient certainty in the value of *End of the Megafauna* for me to think I could successfully take it on. My editor at Norton, John Glusman, was encouraging yet unyielding: a useful combination, as it turned out. I am indebted to all of you.

Photographer Denis Finnin and artists Patricia Wynne and Lorraine Meeker weathered many last-minute requests for images, and always came through. American Museum of Natural History Research Library director Tom Baione and staff members Mai Reitmeyer and Kendra Meyer provided assistance at every stage. At Norton, I also thank editorial assistant Helen Thomaides and copy editor Trent Duffy for dealing with the author's many questions with much patience. For the book's design and careful treatment of both text and imagery, I am grateful to Julia Druskin and Amy Medeiros.

My dear friends and valued colleagues Grant Zazula (Government of Yukon Palaeontology Program, Whitehorse) and Alexandra van der Geer (Naturalis Biodiversity Center, Amsterdam) were willing to read and comment on early drafts. Sometimes I listened; my errors are not theirs.

Finally, and with many fond memories passing across the virtual retina of my mind's eye as I write these lines, I thank my many professional colleagues, from many countries, who have given me such a good run throughout my professional career. These include—but are by no means limited to—Larry Agenbroad (deceased), Tonyo Alcovar, Laura Allen, Lee Arnold, Oscar Arredondo (deceased), Chris Beard, Pere Bover, Alan Bryan (deceased), Bernard Buigues, David Burney, Matt Cartmill, Matt Collins, Joel Cracraft, Bob Dewar (deceased), Denton Ebel, Niles Eldredge, C. W. J. Eliot (deceased), Dan Fisher, Analía Forasiepi, Duane Froese, Don Grayson, Alex Greenwood, Manuel Iturralde-Vinent, Louis Jacobs, Meng Jin, Alejandro Kramarz, Matt Lamanna, George Lyras, Medrie MacPhee,

Preston Marx, Greg McDonald, Donald McFarlane, Malcolm McKenna (deceased), Lorraine Meeker, Carl Mehling, David Meltzer, Dick Mol, Michael Novacek, Pat O'Connor, Hendrik Poinar, Marcelo Reguero, Alberto Reyes, Manuel Rivero de la Calle (deceased), Richard Roberts, Marcelo Sánchez-Villagra, Jeff Saunders, Beth Shapiro, Alfonso Silva-Lee, Sharon Simpson, Graham Slater, David Steadman, Gentry Steele (deceased), Alexi Tikhonov, Sam Turvey, Alexandra van der Geer, Sergey Vartanyan, Martine Vuillaume-Randriamanantena, Neil Wells, Grant Zazula, and, of course, Paul Martin.

CREDITS AND ATTRIBUTIONS

All plates are by Peter Schouten. Credits for text figures are given by figure number, followed by credits for artwork within boxes, given in page order.

Text Figures

Fig. 1.1. Denis Finnin/AMNH

Fig. 1.2. Denis Finnin/AMNH

Fig. 1.3. Denis Finnin/AMNH

Fig. 1.4. Denis Finnin/AMNH

Fig. 2.1. Image by Patricia J. Wynne, after MacPhee (1999), cover art

Fig. 3.1. Image by Patricia J. Wynne, after Zazula et al. (2014), figure 2

Fig. 3.2. Image by Patricia J. Wynne, after Global Warming Art project by Robert A. Rohde

Fig. 3.3. Image by Patricia J. Wynne, after various sources

Fig. 3.4. Image by Patricia J. Wynne, after http://map.ngdc.noaa.gov/website/paleo/paleoclimate/viewer.htm .

Fig. 3.5. Image by Patricia J. Wynne, after http://map.ngdc.noaa.gov/website/paleo/paleoclimate/viewer.htm .

Fig. 3.6. Image by Patricia J. Wynne, after Oppo and Curry (2012), figure 8

Fig. 4.1. Redrawn and updated from MacPhee and Marx (1997)

Fig. 4.2. Image by Patricia J. Wynne, after File:Migraciones austronesias.png by Christophe Cagé ("Map of Expansion of Austronesian Languages," which is based in turn on *Atlas historique des migrations* by Michel Jan et al. [1999] and "The Austronesian Basic Vocabulary Database" [2008])

Fig. 5.1. Image from Cuvier and Brongniart (1822), plate 2A

Fig. 5.2. Image by Patricia J. Wynne

Fig. 5.3. Image from Tilesius (1815), plate 10

Fig. 5.4. Image from Buckland (1824), plate 4

Fig. 5.5. Image from Buckland (1831), appendix, plate 1

Fig. 6.1. Image from Holder (1886), plate facing p. 36

Fig. 6.2. Image by Patricia J. Wynne
Fig. 7.1. Image by Patricia J. Wynne
Fig. 10.1. Image from Martin (1973), figure 2
Fig. 10.2. Image from Waldren and Layard (1872), plate 6
Fig. 10.3. Image from Waterhouse (1839), plate 4
Fig. 11.1. Image by Patricia J. Wynne
Fig. 11.2. Image by Patricia J. Wynne
Fig. 11.3. Image by Patricia J. Wynne
Fig. 12.1. Image by Patricia J. Wynne

Boxes

Page 6, "Mammoth on the Menu." Image from Figuier (1866), Fig. 180
Page 19, "How Fast Can a Species Decline?" Image by Patricia J. Wynne
Page 69, "Cyclops and the Dwarf Elephant" Image by Patricia J. Wynne
Page 73, "Hugest, Fiercest, Strangest! Come See, Come See!" Image from Montulé (1821), unnumbered plate
Page 83, "Wrangel Island, the Woolly Mammoth's Last Stand." Photo by Clare Flemming
Page 87, "Cave Art." Image from Digital Special Collections, AMNH Research Library
Page 121, "*Sporormiella*, the End-Times Fungus." Photo by Clare Flemming
Page 179, "Near Time Extinctions and the Sixth Mass Extinction." Image by Patricia J. Wynne
Page 181, "Reptile Extinctions in Near Time." Image by Patricia J. Wynne

INDEX

Note: Material in illustrations, figures, or tables is indicated by *italic* page numbers. Footnotes or endnotes are indicated by n or nn after the page number.

aardvarks, *61*, *162*
Adams Mammoth, 6, *6*, 71, *71*
Africa
  aridification, 129
  behavioral coevolution, 128–29
  East African Rift panorama, *130–31*
  evidence of hunting by modern humans, 129
  grassland habitat in, 129
  habitats and climate change, 40
  Holocene species, 40–41
  hominin diaspora and, *42*, 44, *44–45*
  Homo sapiens appearance in, 44, 48, *50*
  lack of megafaunal extinctions, 15, 18, 40, 92, 128–29
  during Last Glacial Maximum, 40
  rock art, 40–41
  South African Pleistocene climate, *39*
  South Africa panorama, *38–39*
African elephant, *91*
Agassiz, Louis, 72, 78, 202nn
Akrotiri Aetokremnos, Cyprus, 56
almiquí (*Solenodon cubanus*), *106–7*
Alroy, John, 109, 114
American chestnut tree, 189
American false cheetah (*Miracinoyx trumani*), *37*
American mastodon (*Mammut americanum*)
  Cerutti mastodon site, 52–53, 180
  extinction of, 8, *29*, 199n, 207n
  fossils, *4–5*, 73
  hunting of, 49, 51, 136, 140
  in musth?, *52*
  Page-Ladson kill site, Florida, 49, 51
  as "the American *incognitum*", 73, *73*
American Museum of Natural History, New York City, 4–5, *4–5*, 8, 78, 81, 203n
anak wallaby (*Protemnodon anak*), *122–23*
Ankarana forest, Madagascar, *104–5*
Antarctica temperature, 10,000 to 20,000 BP, *34*
Antarctic Cold Reversal, 208n
Anthropocene extinctions, xii
anthropogenic extinctions, xii
Antillea, or West Indies
  defined, 56
  human arrival in, *42*, 56, 60, 114
  during Last Glacial Maximum, 35
  Near Time extinctions, 24, *42*, 56, 114, *115*, 129
Appalachian Plateau panorama, *110–11*
Arim, Matías, 164
Arroyo del Vizcaíno, Uruguay, 200n
Arroyo Seco 2, Argentina, 201n
Asia
  hominin diaspora and, *42*, 48
  megafaunal extinctions in, 15, 18
  see also Eurasia
Asian elephant, *91*, 188
Asian wild horse (*Equus przewalski*), *153*
Audubon, John James, 19
Australia
  aridification, 124, 182
  climatic variability, 124, 178
  dry woodland panorama, *94–95*
  forest panorama, *122–23*
  human arrival in, *49*, 120, 128, 182
  human–environmental interactions, 127, 178
  hunting in, 124
  lack of radiocarbon dates, 128
  during Last Glacial Maximum, 33
  megafaunal extinction dates, 92, 97, 124, 128
  megafaunal extinctions in, 15, 24, 78, 92, 124, 127
  Naracoorte panorama, *98–99*
  see also Sahul
Austronesian diaspora, *50*, 60

baboon lemur (*Archaeolemur edwardsi*), *58–59*
background rate of extinction, 14
Balouet, Jean-Christophe, 155
bamboo lemur (*Hapalemur griseus*), *13*
bands and surpluses, 143–44, 146–47
baobabs, *58–59*
Barkai, Ran, 203n
bears
  brown bear, 166
  cave bear, 118
  gigantic short-faced bear (*Pararctotherium pamparum*), *22*, *24*
  short-faced bear (*Arctodus simus*), *37*, *167*
  spectacled bear, *24*, *113*
behavioral coevolution, 128–29
Bering land bridge (Beringia)
  ecology, 32, *50*
  exposure in Last Glacial Maximum, *32*
  human migration, 48, *50*, 51

bibymalagasy (*Plesiorycteropus madagascariensis*), *61*
biological first contact, *42–43*, 86, 89, 169, 170
bird extinctions in Near Time, *185*
  see also specific types
Bird, Junius, *135*
bison
  Bison bison, *137*, *139*, 152
  Bison latifrons ("broad forehead"), *137*
  entry into North America, *137*
  Late Pleistocene hunting of, 136, 144
  near extinction of, 203–4n
  predators and, *167*
blitzkrieg
  bow-wave hypothesis, 150–51, *151*, 206
  disease model and, 170
  end of, 89
  human–environmental interactions in Australia, 127
  human interaction long before extinction, 180
  overview, 86, 88, *88*, 108
  simulation models, 89, 183
  speed of extinctions, 136
  targeting of multiple species, 89
  see also Martin, Paul S.; overkill hypothesis
Bluefish Caves, Yukon, 50–51
body size, categorization, 2
Bølling-Allerød, *34*, 186
boreal forests, 28, *29*, 33, 35, *111*, 207n
bow-wave hypothesis, 150–51, *151*, 206
Bradley, Bruce, 51
brown bear (*Ursus arctos*), 166
Buckland, William, *74*, *76*
buff-necked ibis (*Theristicus caudatus*), *vi*
bulldog rat (*Rattus macleari*), *171*
bush meat, 141, 179
bush moa (*Anomalopteryx didiformis*), *62–63*

California, Pleistocene mammals in, *36–37*
camels
  in North America, 35, *36–37*, 139–40, 144, 152
  in South America, 35
  western camel, *36–37*, 152, 207n
Canada, Near Time extinctions in, 24
Canterbury Plains, New Zealand, *62–63*
Cape buffalo (*Syncerus caffer*), *96*
Caral civilization, Peru, 8
carbon-14, 191
  see also radiocarbon dating
Carleton, Michael, 64
catastrophism
  Agassiz's Ice Age concept, 72, 202nn
  biblical scripture and, 70–71, 72, 202n
  Cuvier's geological revolutions, *66–67*, 68, 70–72, 202n
  environmental catastrophism, 68, 70–75
cattle egret (*Bubulcus ibis*), *130–31*
cave art, 87, *87*
cave bear (*Ursus spelaeus*), 118
Cenozoic, dates, 28
Cerutti mastodon site, California, 52–53, 180
Chicxulub impactor, 139, 205
Christmas Island, *171*, 171–72
Christmas Island rat (*Rattus nativitatis*), *171*
Church, George, 189
chytrid fungal infections, 208n
Ciego Montero, Cuba, *57*
climate change
  coevolutionary disequilibrium and, 93, *93*, 97
  explanations of Near Time extinctions, 18, 92–93, 96–97, 102–3, 174–75, 182, 208n
  Late Pleistocene, *29*, 41
  limited effects of, 92, 182, 185–86
  Martin's propositions regarding, 102–3
  Pleistocene, *24*, 28, *29*, *30*
  Pliocene, 28
  Sixth Mass Extinction and, 179
  temperature and ice volume, *30*
  temperature records, *29*, *30*, *34*
  Wallace on glaciations and extinctions, 77–78
cloning, 187–88
Clovis culture
  dates, 109, 139
  end of, 172, 174
  megafaunal overkill and, 139, 140
  stone tools, 51, 136
coevolutionary disequilibrium, 93, *93*, 97
collectors, 206n
Columbian mammoth (*Mammuthus columbianus*), *4–5*, 8, *90–91*, 199n
continental extinctions
  fireball hypothesis, 173–75
  food web collapse hypothesis, 162–66
  hominin diaspora and, 21, 24, 44, 48–53
  hyperdisease hypothesis, 170
  second-order predation hypothesis, 166, 168
  overkill hypothesis, 82, 109–29
  patterns of, 15, 18
  see also island extinctions; *individual continents*
Cooper, Alan, 174
Cordilleran ice sheet, 30, *31*, 48
Cosgrove, Richard, 124
cougar (*Puma concolor*), *84*
CRISPR/Cas9, 188
Cuba, 56, *57*, 114, *115*
Cuban crane (*Grus cubensis*), *57*
Cuban crocodile (*Crocodylus rhombifer*), *57*
Cuban cursorial owl (*Ornimegalonyx oteroi*), *106–7*
Cuddie Springs, Australia, 124
Cuvier, Georges, *67*, 68, 70–72, 78, 202nn
Cuvier's gomphothere (*Cuvieronius hyodon*), *112–13*
Cyclops myth, 69, 201–2n

Dansgaard-Oeschger (D-O) events, *34*, 41
Darwin, Charles
  on extinction, 75, 77, 202n
  influence of Lyell's views, 75
  Macrauchenia patachonica, *145*
  Toxodon platensis, *161*
  warrah or Falkland Islands fox, 155–56, *157*, 207n
Darwin's rhea (*Rhea pennata*), *144–45*
deep-billed megapode (*Megavitiornis altirostris*), *184*
deep time, xii, 75
de-extinction, 187–90
deficient present, 164
Deméré, Tom, 52
Denisovans, 44
developmental cascades, 156, 159
Dewar, Bob, 60
Dinornis novaezelandiae, *148–49*
disease hypothesis, 168–72
disharmony, ecological and evolutionary, 97
DNA studies
  American false cheetah, *84*
  in extinction studies, 108
  mammoths, *91*, 188
  moa, *63*
  Neandertal hybridization with modern humans, *46*
  quagga-coated zebras, *96*

in soils and sediments, 121
Trypanosoma in rats, *171*
dodo (*Raphus cucullatus*), 64
Dolly the Sheep, 187
domestication syndrome, 157, 159
dwarf Cyprus elephant (*Palaeoloxodon cypriotes*), *158*
dwarf Maltese elephant (*Palaeoloxodon falconeri*), *54–55*, *158*
dwarf Maltese hippo (*Hippopotamus melitensis*), *54–55*

East African Rift panorama, *130–31*
Easter Island, *50*
eastern box turtle (*Terrapene carolina carolina*), *181*
ecological or behavioral naïveté, 155–56, 159
ecological release, 86
elephant-foot or thick-set moa (*Pachyornis elephantopus*), *62–63*
elephants
African elephant, *91*
Asian elephant, *91*, 188
dwarf Cyprus elephant (*Palaeoloxodon cypriotes*), *158*
dwarf Maltese elephant (*Palaeoloxodon falconeri*), *54–55*, *158*
extinct savanna elephant (*Loxodonta atlantica*), *38–39*
mammoth-elephant hybrids, 188, 189
poaching of, 141–43
pygmy elephant (*Stegodon florensis*), *47*
Reck's straight-tusked elephant (*Palaeoloxodon recki*), *130–31*
straight-tusked elephant (*Palaeoloxodon namadicus*), *116–17*, 204n
see also American mastodon; mammoths
Escholtz Bay, Alaska, expedition, *76*
Eurasia
hominin diaspora and, *42*, 44, 48, *50*, 114, 118
during Last Glacial Maximum, 32–33, 93
Levant species, *116–17*
megafaunal extinctions in, 18, 24, 114, 118, 120, 182
Eurasian black rats (*Rattus rattus*), 64, *171*, 172, 205n
European ass (*Equus hydruntinus*), *116–17*
Europe and hominin diaspora, *42*, 44, 56, 114
European zebra mussel (*Dreissena polymorpha*), 86, 88
extinct savanna elephant (*Loxodonta atlantica*), *38–39*

facial tumor disease in Tasmanian devils, 208n
facilitated adaptation, 189
Falkland Islands fox or warrah (*Dusicyon australis*), 155–56, *157*, 159, 207n
Fariña, Richard, 164, 166
Fell's Cave, Chile, *135*
Fernando de Noronha, 64
Fiedel, Stuart, 109
Fifth Mass Extinction (K/Pg Mass Extinction), 14, 139, 205–6n
Fijian crocodile (*Volia athollandersoni*), *184*
fireball hypothesis, 172–75, 208n
Firestone, Richard, 172
Fisher, Dan, 144, 146
Flannery, Tim, 127
Flores Island, *18*, *47*, 133, 205n
Font-de-Gaume Grotto, France, *87*
food webs
body size and, 164–66
complexity in Late Pleistocene, 164
disruption hypothesis, 162–66
Late Pleistocene hunters and, 166, 168
second-order predation hypothesis, 166, 168
trophic levels, 164
zebra mussel and, 88
foragers, 206n
four-horned pronghorn (*Captomeryx furcifer*), *36–37*
Frison, George C., 143–44

Gaffney, Eugene, *125*
gene editing tools (CRISPR/Cas9), 188
genetic engineering, 187–90
Geological Evidences for the Antiquity of Man (Lyell), 75, 77
geologic time scale, *3*
giant African buffalo (*Pelorovis antiquus*), *96*
giant anteater (*Myrmecophaga tridactyla*), 8
giant beaver (*Castoroides ohioensis*), *110–11*
giant Cape horse (*Equus capensis*), *38*
giant diprotodont (*Diprotodon optatum*), *94–95*, *98–99*, *162*, 189
giant elephant bird (*Aepyornis maximus*), *xvi–1*, 8, *105*
giant ground sloth (*Lestodon armatus*), *7*, 8, *26–27*, 200n
giant ground sloth (*Megatherium americanum*), *165*, 166, 189
giant Irish deer (*Megaloceros giganteus*), *20*, 120, 182
giant koala lemur (*Megaladapis [Peloriadapis] edwardsi*), 8, *10–11*
giant Malagasy tortoise (*Geochelone grandidieri*), *58–59*
giant Maltese swan (*Cygnus falconeri*), *54–55*
giant narrow-horned hartebeest (*Parmularius angusticornis*), *130–31*
giant primal hartebeest (*Megalotragus priscus*), *38–39*, 41
giant rat of Flores (*Papagomys theodorverhoeveni*), *47*
giant short-faced kangaroo (*Procoptodon goliah*), *94–95*
giant sloth lemur (*Archaeoindris fontoynonti*), 8, *12–13*, *58–59*, 201n
gigantic short-faced bear (*Pararctotherium pamparum*), *22*, *24*
glyptodons
disappearance, 41, 114
key-foot glyptodon (*Glyptodon clavipes*), *vi*
morningstar-tailed glyptodon (*Doedicurus*), *138*, 139
Sclerocalyptus heusseri, *23–24*
in South America, vi, *23–24*, 41, *112*, 114, 139, 143
goanna (*Varanus [Megalania] priscus*), 8, *16–18*, *123*, 181
gomphotheres, 35, 41, 53, *112–13*, 139, 143, 200n
Graham, Russell, 93, 97
Grandidier's koala lemur (*Megaladapis grandidieri*), *104–5*
Grayson, Don, 77
Great American Biotic Interchange, *135*
Greenland, 28, 30, *34*
Greenland ice sheet, *3*, 28, 30
guanaco (*Lama guanicoe*), *112–13*, 139
Guilday, John, 92

Haast's eagle (*Harpagornis moorei*), *176–77*
Harlan's ground sloth (*Paramylodon harlani*), *36–37*
Harlan's musk ox (*Bootherium bombifrons*), *110–11*
harmony, ecological and evolutionary, 97

hartebeest
Alcephalus buselaphus (extant species), *116–17*
giant narrow-horned hartebeest (*Parmularius angusticornis*), *130–31*
giant primal hartebeest (*Megalotragus priscus*), *38–39*, 41
rinderpest epizootic, 170
Haynes, Gary, 139, 140
Heinrich events, *34*, 41, 174
Hemiauchenia paradoxa, *23–24*
hemorrhagic septicemia, 170
henhouse syndrome or surplus killing, 146–47
hippolike diprotodont (*Zygomaturus trilobus*), *98–99*, 122–23, *127*, *162*
hippopotamus
in Britain, *74*
dwarf hippos on Cyprus, 56, 201n
dwarf Maltese hippo, *54–55*
in East African Rift, *130–31*
gorgon hippopotamus, *130–31*
on Madagascar, *58–59*, 60, 201n
Malagasy pygmy hippopotamus, *58–59*, 201n
Malagasy riverine hippopotamus, *58–59*
Hispaniola, 56, 114, *115*
Hispaniolan monkey (*Antillothrix bernensis*), *115*
Holder, Charles Frederick, *80–81*
Holen, Steve, 52
Holocene
climate, 28, 40, 186
dates, *3*
extinctions during, xii, 56, 114, 204n
Homo erectus, *47*
Homo floresiensis ("Hobbit"), *47*, 48, 133
Homo heidelbergensis, 44, *44–45*, 56
Homo neanderthalensis (Neandertals), 44, *46*, 53, *142*
Homo sapiens
appearance in Africa, 44, 48, *50*
interaction with Neandertals, *46*
horse (*Equus sp.*)
Asian wild horse (*Equus przewalski*), *153*
domesticated horse (*Equus caballus*), *153*
in North America, 35, *36–37*, 102, 114, 152, *153*
in South America, 35, *112–13*, *134–35*, 139, *153*
in southern Africa, 38–39
tarpan or wild horse, *153*
human population of Earth, 44
hunter-gatherers, 82, 85, 128, 143–44, 146, 206n
Hutton, James, 72
hyenas, 72, *74*, *116–17*, 146
hyperdisease hypothesis, 168–72, *169*

Ice Age, 28–41
Agassiz's concept of, 72, 202nn
effects on sea level and coastlines, 32, *32*
ice sheets in Northern Hemisphere, 30, *31*, 32
vegetation zones during, 32, *33*, 35
see also Last Glacial Maximum
Ice Age extinctions, xii, 75, 202n
see also Near Time extinctions
ice ages
in Pliocene, 28
temperature and ice volume, *30*
see also Ice Age; Last Glacial Maximum
Indian darter (*Anhinga melanogaster vulsini*), *58–59*
Indonesia
Flores Island, *18*, *47*, 133, 205n
human arrival in, *47*, 48
during Last Glacial Maximum, 33
megafaunal extinctions, *132*, 133
Pleistocene fauna, *132*
Innuitan ice sheet, 30, *31*
interglaciations
Last Interglaciation, *3*, *29*, 180, 207n
in Pliocene, 28
interstadial events, *34*, 175
invasive or introduced species
in Antillea, 129, 133
brown tree snakes on Guam, 205n
dingoes and foxes in Australia, 147
ecological release, 86
Eurasian black rats on Christmas Island, *171*, 172
on Mauritius, 64
in New Zealand, *149*, *179*, 201n
invisible extirpation, 108
island extinctions
bird extinctions, *185*
hominin diaspora and, 24, 56, 60–64, 129, 133
hyperdisease hypothesis, 171–72
islands as refugia, 33
speed of extinctions on, *43*
see also continental extinctions; *individual islands*
island syndrome, 152, 155–57

jaguar (*Panthera onca*), *84*, *110–11*
Jamaica, 56, 60
Java, 44, *132*
Javanese dwarf stegodon (*Stegodon hypsilophus*), *132*, 133
Jebel Irhoud, Morocco, 48, *96*
Jefferson's ground sloth (*Megalonyx jeffersonii*), *76*, *110–11*, 207n
Juma giraffe (*Giraffa jumae*), *130–31*

kakapo (*Strigops habroptila*), *179*
Kennett, James, 172
key-foot glyptodon (*Glyptodon clavipes*), *vi*
"keystone herbivore" hypothesis, 204n
kill sites
lack of, 136–40, 183, 185
Page-Ladson site, Florida, 49, 51, 140
Kirkland Cave, Yorkshire, England, *74*
Knight, Charles R., *87*
Kolyma basin, Russia, 189
Komodo dragon (*Varanus komodensis*), 15, *18*
kookaburra (*Dacelo novaeguineae*), *94–95*
K/Pg Mass Extinction, 14, 139, 205–6n
Kruuk, Hans, 146–47

Lakaton'i Anja, Madagascar, 60
Last Glacial Maximum
dates, *3*, *29*, 30, *50*
effects on sea level and coastlines, 32, *32*
ice sheets in Northern Hemisphere, 30, *31*, 32
negligible losses in Northern Hemisphere, 186
vegetation zones during, 32, *33*, 35
Last Interglaciation (Last Interglacial), *3*, *29*, 180, 207n
Late Pleistocene
climate, *29*, 41, 186
dates, *3*
extinctions in, xii, 97, 102, 139, 186
see also Near Time extinctions
late Quaternary extinctions, xii, 120, 182
see also Near Time extinctions
Laurentide ice sheet, 30, *31*, 35, 48

lemurs
baboon lemur, *58–59*
bamboo lemur, *13*
disappearance in Madagascar, 103
giant koala lemur, 8, *10–11*
giant sloth lemur, 8, *12–13*, *58–59*, 201n
Grandidier's koala lemur, *104–5*
long-armed sloth lemur, *104–5*
ringtailed lemur, *xvi–1*
tretretretre sloth lemur, *13*, *104–5*
see also Madagascar
Levant species, *116–17*
Liang Bua site, Flores Island, *47*, 133, 205n
Little Ice Age, *3*, 41
llamas, *23–24*, *36–37*
Llamas, Bastien, 51
long-armed sloth lemur (*Mesopropithecus dolichobrachion*), *104–5*
long-nosed peccary (*Mylohyus nasutus*), *110–11*
Lord Howe Island, *125*, 155
Lord Howe pigeon (*Columba vitiensis godmanae*), *154*, 155, 159
Lundelius, Ernest, 93, 97
Lyell, Charles, 72, 75, 77

Macrauchenia patachonica, *144–45*, *161*
Madagascar
Ankarana forest, *104–5*
bibymalagasy, *61*
elephant birds, *1*, *58–59*
giant Malagasy tortoise, *58–59*
highlands panorama, *58–59*
human arrival in, *42*, *50*, 60, 201n
Indian darter, *58–59*
landscape alteration from burning, 82, 203n
Madagascar crocodile, *58–59*
Malagasy pygmy hippopotamus, *58–59*, 201n
Malagasy riverine hippopotamus, *58–59*
Near Time extinctions, 24, 60, 183
see also lemurs
Madagascar crocodile (*Voay robustus*), *58–59*
Madjedbebe, Australia, 120, 124
Malagasy pygmy hippopotamus (*Hexaprotodon madagascariensis*), *58–59*, 201n
Malagasy riverine hippopotamus (*Hippopotamus lemerlei*), *58–59*
Malta panorama, *54–55*
mammoths
cloning and, 187–88
Columbian mammoth, *4–5*, 8, *90–91*, 199n
eating of, 6, 199n
ecological niche, 97, 102
extinction, 8, 71–72, 103, 199n
ivory scoop from a fossil tusk, *76*
mammoth-elephant hybrids, 188, 189
mammoth hunting, *80–81*, 118, 136, 140, 142–43
tundra mummies, 6
on Wrangel Island, 83, *83*, 120
see also woolly mammoth
mammoth steppe, 32, 40, *100–102*, 102, 189
maned wolf (*Chrysocyon brachyurus*), *157*
Maoris, 60, *63*, *149*, *177*, *179*, 201n
marsh deer (*Blastocerus dichotomus*), *165*
marsupial "panda" (*Hulitherium tomasetti*), *126–27*
marsupial "tapir" (*Palorchestes azael*), *122–23*, *162–63*
Martin, Paul S.
behavioral coevolution, 128–29
bow-wave hypothesis, 150–51, *151*, 206
correlation between human arrival and species disappearances, 24, 82, 103, 109, 120, 180
on fast extinction in Madagascar, 60, 82
global explanation for extinctions, 92, 102
human arrival date in New World, 48
on losses of largest animals, 21, 24
prey naïveté, 21, 85–86, 143, 150, 155–56
propositions on climate change, 102–3
radiocarbon dating used by, 79, 82, 108, 136, 203n
rewilding, 189
simulation models of faunal extinction, 89, 109, 114
on surplus killing, 147
syncopation scenario, 103, 109, 120, 140
see also blitzkrieg; overkill hypothesis
Marx, Preston, 168, 170
mastodons. *see* American mastodon
Mauritius, *42*, 64
Mayor, Adrienne, 201–2n
Mediterranea
definition and islands in, *42*, 56
human arrival in, *42*, 56
Malta panorama, *54–55*
megafaunal extinctions
alternate names, xii
dates, 9
distribution in time and space, *15*
number of species lost, 21, 24
pattern features of, 14–15, 18
premodern hominins not associated with, 48
reproductive and maturation rates and, 14–15, 103, 180
see also Near Time extinctions; overkill hypothesis
megafaunal species, defined, 2n, *15*
Megalocnus rodens, *57*
megatherium (*Megatherium tarijensis*), *112–13*, 139
meiolaniid turtles, *125*, 181
mekosuchian crocodiles, *123*, *184*
Merck's rhinoceros (*Stephanorhinus hemitoechus*), *116–17*, 118, 120, *142*, 206n
meridiungulates, 35, 143, *144–45*, *160–61*, 200n
metallic pigeon (*Columba vitiensis griseogularis*), *154*, 155
Metcalf, Jessica, 174
Middle Kingdom, Egypt, 8
mihirung (*Genyornis newtoni*), *94–95*, *98*
Milankovitch cycles, 28
moa, 60–61, *62–63*, *148–49*, *177*
modern-era extinctions, xii
see also Near Time extinctions
Monte Verde, Chile, 53, 201n, 208n
morningstar-tailed glyptodon (*Doedicurus*), *138*, 139
Mossiman, James, 89
Mucuñuque, Venezuela, 173
Muller's large elephant bird (*Mullerornis grandis*), *1*, *58–59*, *104–5*
musk ox, *76*, *110–11*, 182

Naracoorte, South Australia, *17–18*, *98–99*
Near Time extinctions
bird extinctions, *185*
climate change arguments, 18, 28, 92–93, 96–97, 102–3, 174–75, 208n
common-cause explanations, 102, 118, 182

Near Time extinctions (*continued*)
dates, xii, 15
defined, xii
distribution in time and space, 14, *15*
early attempts to explain, 66–79
fireball hypothesis, 172–75, 208n
food web disruption and, 162–68
future research directions, 183–86
Holocene extinctions, xii, 56, 114, 204n
human–environmental interactions and, x–xii, 18–19, 78–79, 127, 178–79, 186
hyperdisease hypothesis, 168–72, *169*
largest species most affected by, 14–15, 18
number of species lost, 21, 24
pattern features of, 14–15, 18
reptile extinctions, 181
Sixth Mass Extinction and, 178, 179
of terrestrial vertebrates, 14
see also megafaunal extinctions; overkill hypothesis
neural crest, 156
New Guinea, 33, 120, 124, *126–27*, 127–28, *154*
New Zealand
Austronesian diaspora, *50*, 60
Canterbury Plains panorama, *62–63*
evidence of hunting by modern humans, 60–61, 201n
human arrival in, *50*, 60, 201n
moa, 60–61, *62–63*, *148–49*, *177*
New Zealand wrens (*Pachyplichas yaldwyni*), *148–49*
ninja turtle (*Ninjemys oweni*), *94–95*, *125*
Noronhomys vespuccii, 64
North America
Appalachian Plateau panorama, *110–11*
bow-wave hypothesis, *151*
early human presence in, *42*, 48–53, *50*, 114, 180
evidence of megafaunal hunting, 136, 140
human migration down west coast, *50*, 51
human migration, Solutrean hypothesis, 51
human migration through ice-free corridor, 48–50, 89, 200n
lack of kill sites, 136, 139
during Last Glacial Maximum, 30, 33, 35, 49–50, 93, 180
megafaunal extinctions in, 15, 21, 24, 109, 114, 199n
megafaunal species and fossils, 35, *36–37*, 40
oldest dated human remains, 53
Southern California panorama, *36–37*
speed of megafaunal extinctions in, 109, 114, 143
northern white rhino, 206n
North Island giant moa (*Dinornis novaezelandiae*), *148–49*

Olson, Storrs, 64
o'o, *185*
optically stimulated luminescence (OSL), 53, 97, 128, 192
orangutans, *13*
Osborn, Henry Fairfield, 78–79
OSL (optically stimulated luminescence), 53, 97, 128, 192
ouakialoa, *185*
overkill hypothesis
bands and surpluses, 143–44, 146–47
Clovis culture and, 139, 140
as common-cause explanation for Near Time extinctions, 118, 182–83
connotations of overhunting, 140–41
correlation between human arrival and species disappearances, 24, 82, 103, 109, 120, 180
current state of, 108–33, 180, 182
early explanations of extinctions, 75–79
food security and, 88, 141, 203n
henhouse syndrome or surplus killing, 146–47
lack of kill sites, 136–40, 183, 185
Lyell's view on, 75, 77
objections to, 92, 135–47, 150–59, 178, 180, 182
overview, Martin's overkill hypothesis, 24–25, 80–89, 178
paleoeconomics of overhunting, 140–44, 146–47
parallels to rhino and elephant poaching, 141–43
passenger pigeons, 19
simulation models of faunal extinction, 89, 109, 114, 183
see also blitzkrieg
Owen, Richard, *63*, 78
Owen-Smith, Norman, 204n
oxygen isotope data, *34*

Page-Ladson kill site, Florida, 49, 51, 140
paleogenomics, 183
Panamanian isthmus, 28, 35, *135*
páramo, 35, *112–13*
Paris basin, *66–67*
Parocnus serus, *115*
passenger pigeon (*Ectopistes migratorius*), 19, *19*, 189, 199n, 209n
Patagonian ice sheet, 35, *113*
Paula's tree kangaroo (*Bohra paulae*), *122–23*
Peale, Charles Willson, 73
Peale, Rembrandt, 73
Persian fallow deer (*Dama dama mesopotamica*), *158*
petroglyphs, 87
pictographs, 87
pink cockatoo (*Cacatua leadbeateri*), *94–95*
Pinta Island tortoise (*Chelonoidis abingdonii*), *181*
Pitulko, Vladimir, 114, 118
plains zebra (*Equus quagga*), *38–39*, *96*, *130–31*
Pleistocene climate, *24*, 28, *29*, *30*, 186
Pleistocene dates, *3*
Pleistocene–Holocene transition, 41, 97, 132, 152, 191
Pliocene, *3*, 28, 124, *153*
Polo, Marco, *1*
Polynesian or kiore rat (*Rattus exulans*), *149*, 201n
Polyphemus the Cyclops, *69*
predator-prey relationships, 85–86, 203n
prey naïveté, 21, 85–86, 143, 150, 155–56, 159
proboscidean, defined, 8
pronghorn (*Tetrameryx*), *84*
Puerto Rico, 56, 114
pygmy elephant (*Stegodon florensis*), *47*

quagga coloration of plains zebra, *38–39*, *96*
Quammen, David, 155
Quaternary dates, *3*, 14
Quaternary, temperature and ice volume in, *30*

radiocarbon dating
human arrival in New Zealand, 60
invention and improvements, 79, 108

mammoths on Wrangel Island, 83, *83*
Martin's use of, 79, 82, 108, 136, 203n
North American sites, 49, 51, 114, 191
overview and calibration, 191
Sahul, lack of good dating in, 128
South American archeological sites, 53
Ramsay's echidna or spiny anteater (*Megalibgwilia ramsayi*), *98–99*
Reck's straight-tusked elephant (*Palaeoloxodon recki*), *130–31*
red kangaroo (*Osphranter rufus*), *95*, 120
red-necked wallaby (*Macropus rufogriseus*), *98–99*
reptile extinctions in Near Time, 181
see also specific types
rewilding, 189–90
rhinos
Merck's rhinoceros, *116–17*, 118, 120, *142*, 206n
northern white rhino, 206n
poaching of, 141–43
steppe rhino, *119*
Sumatran rhino, *119*
white rhino, *38–39*, 206n
woolly rhino, *118–19*, 204n
ribbon-tailed bird of paradise (*Astrapia mayeri*), *126–27*
rinderpest epizootic, 170
ring-necked pheasant (*Phasianus colchicus*), *20*
ringtailed lemur (*Lemur catta*), *xvi–1*
rock pigeon, 189
Roc, legend of, *1*
roe deer (*Capreolus capreolus*), *46*
Romer, Alfred S., 78, 79, 82
rufous oven bird (*Funarius rufus*), *vi*

sabertooth cat (*Smilodon fatalis*), *ii*, 8, *9*, *37*, *167*, 189
sable antelope (*Hippotragus niger*), *38–39*
Sahul
human arrival in, *42*, 44, 48, *49*
lack of radiocarbon dates, 128
megafaunal extinctions in, 120, 124, 127–28
see also specific locations
saiga antelope (*Saiga tatarica*), *119*, 170
Saltré, Frédérik, 127–28
Sánchez-Villagra, Marcelo, 156
Schoneningen, Germany, 206n
Schouten, Peter, xiii
science, process of, 72, 73
Sclerocalyptus heusseri, *23–24*
sclerophyllous scrub, 124, 127
sea level in Last Glacial Maximum, 32, *32*
second-order predation hypothesis, 166, 168
Segura, Angel, 164
Shasta ground sloth (*Nothrotheriops shastensis*), *36*
short-faced bear (*Arctodus simus*), *37*, *167*
short-faced kangaroo, giant (*Procoptodon goliah*), *94–95*
short-faced kangaroo (*Procoptodon raphe*), *125*
short-faced kangaroo (*Simosthenurus occidentalis*), *95*, *98–99*
Siberia
Late Pleistocene mammoth hunting, 114, 118
Near Time extinctions, 24
Simmons, Alan, 56, 201n
sivathere (*Sivatherium maurusium*), *130–31*
Sixth Mass Extinction, 178, 179
Slavenko, Alex, 181
slender-headed mylodont sloth (*Scelidotherium leptocephalum*), *112–13*
sloths
in Arctic North America, *76*
disappearance of ground sloths, 41, 114
giant ground sloth (*Lestodon armatus*), *7*, 8, *26–27*, 200n
giant ground sloth (*Megatherium americanum*), *165*, 166, 189
Harlan's ground sloth, *36–37*
Jefferson's ground sloth, *76*, *110–11*, 207n
Megalocnus rodens, *57*
megatherium (*Megatherium tarijensis*), *112–13*, 139
Parocnus serus, *115*
Shasta ground sloth, *36*
slender-headed mylodont sloth, *112–13*
three-toed tree sloths, 8, *165*
two-toed tree sloths, 8
smooth-billed ani (*Crotophaga ani*), *26–27*
Solutrean hypothesis, 51
Song of the Dodo, The (Quammen), 155
Sopochnaya Karga, Siberia, 114, 118
South America
aridification, 166
bow-wave hypothesis, *151*
evidence of megafaunal hunting, 139
food web disruption, 164–66
human arrival in, *42*, 53, 200–201nn, 208n
during Last Glacial Maximum, 35
megafaunal extinctions in, 15, 21, 41, 208n
megafaunal species and fossils, 35, 40
páramo, 35, *112–13*
vegetation zones, 35
South Island giant moa (*Dinornis robustus*), *62*
spectacled bear (*Tremarctos ornatus*), *24*, *113*
Sporormiella, 121, 124
spotted hyena (*Crocuta crocuta*), *74*, *116–17*, 146
stag moose (*Cervalces scotti*), *110–11*
Stanford, Dennis, 51
Steadman, David, 147
steppe rhino (*Elasmotherium sibiricum*), *119*
stilt-legged llama (*Hemiauchenia macrocephala*), *36–37*
straight-tusked elephant (*Palaeoloxodon namadicus*), *116–17*, 204n
strong-beaked crocodile (*Quinkana fortirostrum*), *122–23*
Sumatran rhino (*Dicerorhinus sumatrensis*), *119*
Sundaland, 48, *132*
surrogate mothers and genetic engineering, 188
synthetic biology, 187–90

taiga, 33, 189
tameness, 152, *154*, 155–56, 159, 186
Taolambiby, Madagascar, 60
tarpan or wild horse, *153*
Tasmanian devil (*Sarcophilus harrisii laniarius*), *98–99*, 208n
Teenage Mutant Ninja Turtles, *125*
thermohaline circulation in oceans, 28, 124
three-toed tree sloths (*Bradypus*), 8, *165*
thylacine or marsupial "tiger" (*Thylacinus cynocephalus*), *98–99*, 208n
thylacoleo or marsupial "lion" (*Thylacoleo carnifex*), *98–99*, *122–23*
Tikhonov, Alexei, 118
Toxodon platensis, *160–61*, 204n

tretretretre sloth lemur (*Palaeopropithecus maximus*), *13*, *104–5*
trumpeter swan (*Cygnus buccinator*), *55*
trypanosomiasis, *171*, 172
Twilight of the Mammoths (Martin), 89
two-toed tree sloths (*Choloepus*), 8

uniformitarianism, 72, 75

Vartanyan, Sergey, 83
vero tapir (*Tapirus veroensis*), *110–11*
Vespucci, Amerigo, 64

Wakefield's wombat (*Warendja wakefieldi*), *98–99*
wallabies, *16–18*, *98–99*, *122–23*, 124
Wallace, Alfred Russel, vii, 14, 77–78, 202–3n
Wally's Beach, Alberta, Canada, 140
warrah or Falkland Islands fox (*Dusicyon australis*), 155–56, *157*, 159, 207n
warthog (*Metridiochoerus modestus*), *44–45*
warthog (*Phacochoerus africanus*), *44*
water buffalo (*Bubalus palaeokerabau*), *132*
western camel (*Camelops hesternus*), *36–37*, 152, 207n
West Indies, 24
white rhino (*Ceratotherium simus*), *38–39*, 206n
Whitney-Smith, Elin, 166
wild Asian water buffalo (*Bubalus arnee*), *132*
Wilmot, Ian, 187
wonambi (*Wonambi naracoortensis*), 49, *94–95*, 181
woolly mammoth (*Mammuthus primigenius*)
  Adams Mammoth, 6, *6*, 71, *71*
  Cuvier's mechanism for disappearance, 71–72
  extinction, 8, 71–72, 118, 182, 199n
  herd on Siberian steppe, *100–102*
  hybridization with Columbian mammoths, *91*
  on St. Paul Island, Alaska, 120
  on Wrangel Island, 83, *83*, 120, 203n
woolly rhino (*Coelodonta antiquitatis*), *118–19*, 204n
Wrangel Island, *42*, 83, 120, 203n
Wroe, Steve, 182
Wynne, Patricia, xiii, *19*

Xenarthra, 8

Younger Dryas
  bolide or cometary impactor and, 172–73
  cold interval, *34*, 40, 92, 174, 186, 207n
  dates, *3*, 207n

zebra mussel (*Dreissena polymorpha*), 86, 88
Zimov, Sergey, 189